AF581707

High-Efficiency Crystalline Silicon Solar Cells

High-Efficiency Crystalline Silicon Solar Cells

Editors

Eun-Chel Cho

Hae-Seok Lee

MDPI • Basel • Beijing • Wuhan • Barcelona • Belgrade • Manchester • Tokyo • Cluj • Tianjin

Editors
Eun-Chel Cho
Sungkyunkwan University
Korea

Hae-Seok Lee
Korea University
Korea

Editorial Office
MDPI
St. Alban-Anlage 66
4052 Basel, Switzerland

This is a reprint of articles from the Special Issue published online in the open access journal *Energies* (ISSN 1996-1073) (available at: https://www.mdpi.com/journal/energies/special_issues/Crystalline_Silicon_Solar_Cells).

For citation purposes, cite each article independently as indicated on the article page online and as indicated below:

LastName, A.A.; LastName, B.B.; LastName, C.C. Article Title. *Journal Name* **Year**, *Article Number*, Page Range.

ISBN 978-3-03943-629-3 (Pbk)
ISBN 978-3-03943-630-9 (PDF)

Contents

About the Editors

Eun-Chel Cho (Research Professor) received his Ph.D. in Photovoltaics from the University of New South Wales (UNSW), Australia in 2003. He joined the Samsung Advanced Institute of Technology as a member of research staff in 1993, Samsung SDI in 2004, and UNSW as a Research Fellow in 2006. He joined Hyundai Heavy Industries in 2008 and served as the Senior vice president and Director of the Green Energy Research Institute at Hyundai Heavy Industries Co., Ltd. He has conducted research on silicon quantum dot research in UNSW and device improvement of commercialization of industrial PERC (Passivated Emitter Rear Contact) solar cells in Hyundai Heavy Industries. His research interest is mainly focused on the industrialization of highly efficient crystalline silicon solar cells (p-type & n-type) and module technology including life cycle reliability and power generation at the system level. Dr. Eun-Chel Cho is currently Research Professor at Sungkyunkwan University. He has authored over 90 publications in international journals and over 40 patents for PV. His total number of citations is over 3500.

Hae-Seok Lee (Professor) received his Ph.D. degree from the National Toyohashi University of Technology (TUT) in 2003, where his research focused on $CuInSe_2$ (CIS) space solar cells. From 2003 to 2006, he worked as a research fellow at Toyota Technological Institute, where he studied super high-efficiency InGaP/InGaAs/Ge multi-junction solar cells and concentrator systems ($\eta \sim 38.9\%$). In March 2006, he came back to Korea and worked as a chief researcher at LG Electronics to develop high-efficiency large-area amorphous Si-based thin-film solar cells. From July 2008 to February 2013, he joined Shinsung Solar Energy Co., Ltd. as a Managing Director in the mass production of crystalline Si(c-Si) solar cells. Since 2013, he has been with Korea University, where his research interest covers the development of high efficiency c-Si solar cells & modules, perovskite/silicon tandem solar cells, and nano-materials for solar cells. He is currently leading on the development of 35% efficiency "Super solar cell, perovskite/silicon tandem" as a national alchemist project. He has authored over 100 publications in international journals and over 50 patents for PV. His total number of citations is over 3000.

Preface to "High-Efficiency Crystalline Silicon Solar Cells"

Among green energy resources, solar photovoltaics are in huge demand worldwide, as they are now used to supply electrical power and are considered a good replacement for fossil fuel. Among the many photovoltaic devices, crystalline silicon solar cells and their system application occupy more than 95% of the photovoltaic market. High-efficiency cell structures help to reduce the costs of photovoltaic energy generation in two ways: (i) by increasing the efficiency and, hence, the power output per area of used silicon or (ii) by allowing the use of thinner wafers, achieving the same level or even improved efficiency and, hence, the power output per volume or weight. However, four important aspects are associated with high-efficiency crystalline silicon solar cells, that is, (i) the surface passivation, (ii) metal contacts, (iii) material quality, and (iv) cell structure.

Hence, this Special Issue focuses on contributions on high-efficiency crystalline silicon solar cells with enhanced scientific and multidisciplinary knowledge to improve performance and deployment for PV energy security. In this book, the features of the high-efficiency crystalline silicon solar cells such.

Eun-Chel Cho, Hae-Seok Lee
Editors

Article

A Novel Method to Achieve Selective Emitter Using Surface Morphology for PERC Silicon Solar Cells

Minkyu Ju [1], Jeongeun Park [1], Young Hyun Cho [2], Youngkuk Kim [2], Donggun Lim [1], Eun-Chel Cho [2,*] and Junsin Yi [2,*]

1 School of Electronic, Electrical Engineering, Korea National University of Transportation, Chungju 27469, Korea; mkju@ut.ac.kr (M.J.); ac1331@ut.ac.kr (J.P.); dglim@ut.ac.kr (D.L.)

2 School of Information and Communication Engineering, Sungkyunkwan University, Suwon 16419, Korea; yhcho64@skku.edu (Y.H.C.); bri3tain@skku.edu (Y.K.)

* Correspondence: echo0211@skku.edu (E.-C.C.); junsin@skku.edu (J.Y.)

Received: 24 August 2020; Accepted: 29 September 2020; Published: 6 October 2020

Abstract: Recently, selective emitter (SE) technology has attracted renewed attention in the Si solar cell industry to achieve an improved conversion efficiency of passivated-emitter rear-contact (PERC) cells. In this study, we presented a novel technique for the SE formation by controlling the surface morphology of Si wafers. SEs were formed simultaneously, that is, in a single step for the doping process on different surface morphologies, nano/micro-surfaces, which were formed during the texturing processes; in the same doping process, the nano- and micro-structured areas showed different sheet resistances. In addition, the difference in sheet resistance between the heavily doped and shallow emitters could be controlled from almost 0 to 60 Ω/sq by changing the doping process conditions, pre-deposition and driving time, and temperature. Regarding cell fabrication, wafers simultaneously doped in the same tube were used. The sheet resistance of the homogeneously doped-on standard micro-pyramid surface was approximately 82 Ω/sq, and those of the selectively formed nano/micro-surfaces doped on were on 62 and 82 Ω/sq, respectively. As a result, regarding doped-on selectively formed nano/micro-surfaces, SE cells showed a J_{SC} increase (0.44 mA/cm^2) and a fill factor (FF) increase (0.6%) with respect to the homogeneously doped cells on the micro-pyramid surface, resulting in about 0.27% enhanced conversion efficiency.

Keywords: selective emitter; surface morphology; doping process; PERC; solar cell

1. Introduction

Presently, crystalline silicon (c-Si) solar cells are the leading product in the solar cell market, owing to their efficiency and low production cost [1,2], and this trend is expected to continue [3]. In terms of cell fabrication technology, passivated-emitter rear-contact (PERC) cells have become the mainstream product in the mass production of c-Si solar cells, owing to their high efficiency and cost-effectiveness [4–6]. According to an ITRPV (International Technology Roadmap for Photovoltaics) report, the market share of PERC technology in 2019 was over 30% and, according to estimates, it will exceed 60% by 2030 [3]. With continuous development over the past few years, PERC cell technology has already achieved a mass-production efficiency exceeding 21% [6]. Therefore, to further improve the efficiency, most solar cell companies are trying to develop new technologies and apply them to their production lines. The selective emitter (SE) technology has recently again attracted renewed attention, owing to its high efficiency [3–10].

SE technology has the following advantages. First, it has low sheet resistance because the high doping concentration under the electrodes lowers the contact resistance, leading to an improvement in the fill factor (FF). Second, better surface passivation on the high-sheet-resistance area enhances the short-circuit current density (J_{SC}) and open-circuit voltage (V_{OC}) [6–10]. There are several methods to create an SE, such as dopant-paste printing [11,12], inkjet printing [13,14], etch-back [15,16], and laser techniques [17–24]. However, most of these techniques are applied immediately before or after the doping process, so there are risks to the solar cell manufacturing process. As shown in Figure 1, SE technologies such as dopant-paste printing and inkjet printing are susceptible to contamination; in other words, there is a risk of contaminants spreading into the silicon during the high-temperature doping process [11–14]. Etch-back and etch-paste techniques result in the loss of reflectivity due to smoothing of the texture surfaces [15–17]. The laser doping method can damage the silicon surface, which is detrimental to the open-circuit voltage (V_{OC}) [18,19]. In addition, most SE technologies are likely to cause misalignment, which results in over-alignment losses in the alignment with the front electrode during the printing process [8–23]. Recently, as a new technology to overcome this, the development of a new SE to mitigate the recombination loss by the heavily doped epitaxial layer selectively grown after the passivation process, has been reported [24].

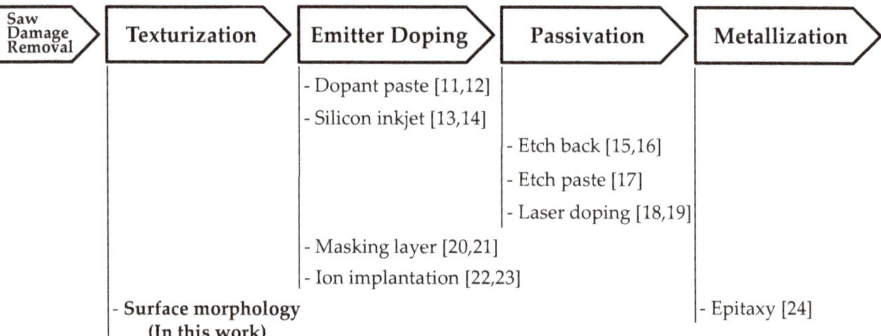

Figure 1. Comparison between the processes for selective emitter (SE) technologies used in our study and in previous studies [11–24].

In this study, a novel SE technique that uses selective nanosurface morphologies for a heavy emitter area was proposed before the pyramid texturing process [3–24]. A new SE technique that uses each different surface morphology was proposed. Even though the nano/micro-structured areas were doped in a single doping process, they exhibited different sheet resistances. In addition, the differences in sheet resistance could be controlled during the doping process by altering the temperature, pre-deposition time, and drive-in time. Importantly, using this SE technique could perform a sufficient cleaning process before the doping process, so there was less contamination risk to the solar cell manufacturing process. Furthermore, the loss from the mismatch between the heavily doped region and the front electrodes by screen-printing could be drastically reduced. Therefore, compared to conventional SE, there was less risk such as surface damage, return loss, and over-alignment loss.

2. Experimental Details

In this experiment, p-type boron-doped 6-inch single c-Si wafers with a resistivity of 1–1.5 ohm-cm and a thickness of 180 ± 20 μm were used. Initially, the c-Si wafers were immersed for 10 min at approximately 75 °C, in a mixed etching solution with a concentration of 5 wt% sodium hydroxide (NaOH) and 0.75 wt% sodium hypochlorite (NaOCl), to ensure saw damage removal (SDR). The wafers were rinsed in flowing DI water, followed by hydrochloric acid (HCl) and hydrofluoric acid (HF), to remove the metal ions and native oxides on the surface. Finally, the wafers were rinsed again in the flowing DI water and then dried.

2.1. Characteristics of Phosphorus Doping According to Nanosurface Morphology

To obtain the nanostructure on one side, two cleaned SDR wafers in contact with each other were placed on the etching carrier. The nanostructure texture was obtained through vapor-texturing using an etching solution of HF:HNO_3 = 7:3 volume ratio to form a uniform nanostructure on the surface [25]. The morphologies of the samples were controlled, and the shape and size of the surface nanostructure were modified during etching by varying the post-etching time between 0 and 3 min in an HF:HNO_3:CH_3COOH isotropic etching solution with a volume ratio of (1:100:50), at room temperature. After vapor texturing, the samples were immersed for 10 min at 75 °C in the standard clean 1 (SC1) solution of NH_4OH:H_2O_2:H_2O (1:1:5 volume ratio) to remove the chemicals remaining on the surface. Then, the samples were rinsed in flowing DI water followed by HF, to remove the native oxides on the surface. Finally, the wafers were rinsed again in flowing DI water and were dried. To form the micro-pyramid structure on the other surface of the samples, a 160-nm thickness silicon nitride (SiN_X) texturing barrier was deposited on the nanosurface using plasma-enhanced chemical vapor deposition (PECVD). Then, the remaining silicon particles on the surface were removed from the samples through SC1 cleaning. The samples were immersed for 30 min at 81–83 °C in a mixed solution of 2 wt% sodium hydroxide (NaOH), 6.25 wt% sodium silicate (Na_2SiO_3), and 12.5 vol% isopropyl alcohol (IPA) for micro-pyramid texturization [26]. For removal of the SiN_X texture-barrier film formed on the nanostructure, the samples were immersed for 10 min at room temperature in buffered HF (BHF, NH_4F:HF = 6:1). The samples were rinsed in flowing DI water, followed by HCl and HF, to remove the metal ions and native oxides on the surface. Finally, the wafers were rinsed again in flowing DI water and dried.

Phosphorus doping was performed at a temperature of 825 °C for 10 min in a quartz-tube furnace using $POCl_3$ for the formation of an emitter layer on the surface of the samples, with each sample having a different surface. To remove the phosphorus silicate glass (PSG) layer deposited on the surface, the sample was immersed in an HF solution for 30 s, and then washed with DI water and dried. The sheet resistances of the emitters formed according to the varied surface structure were measured at similar positions, at a uniform distance apart, through a four-point probe. The surface morphology of samples with varied nanostructure was observed using scanning electron microscope (SEM) images.

2.2. Characteristics of Nanosurface Emitter Layer According to Doping Condition

To observe the variation in sheet resistance according to the phosphorus doping condition in the variable surface morphology, wafers formed on both sides with a nanostructure and a micro-pyramid structure were used. The prepared samples were cleaned with HCl and HF, rinsed with flowing DI water, and dried. The cleaned samples were processed in a quartz-tube furnace using $POCl_3$ at a temperature range of 805–860 °C for 10 min, to ensure a varied sheet resistance for the shallow emitter on the surface. Subsequently, the samples were processed in a quartz-tube furnace using $POCl_3$ for 10–30 min at 860 °C, to ensure a varied sheet resistance for the deep emitter on the surface. After removing the PSG layer by using HF, the sheet resistances of the emitter formed on the nano- and micro-pyramid structures were measured and analyzed. The analysis of the number of phosphorus atoms in the emitter with varied surface morphology was performed using a CAMECA IMS 7f magnetic-sector secondary ion mass spectroscope (SIMS) (CAMECA in Gennevilliers, France).

2.3. Fabrication of Novel SE Solar Cells Using Surface Morphology

To analyze the characteristics of the solar cells according to their surface morphology, they were designed to have a nanostructure (nano), micro-pyramid structure (pyramid), and nano-micro-SE structure (selective). Initially, all the fabricated samples had a uniform nanostructure on the front side from the cleaned SDR by vapor-texturing [25]. For the sample with the nano-micro-SE structure, a SiN_X texturing barrier was deposited on the front side using PECVD. The front texture barrier was patterned using the acid etching resistance through screen-printing using a front-grid mask with a line width of about 100 μm, and spacing of 1.312 mm. Then, the sample was immersed for 10 min at room temperature, and BHF was used to pattern the surface by removing the SiN_X texture barrier. All the samples were immersed for 10 min at 75 °C in an SC1 solution to remove the remaining chemicals on the surface. Then, the acid etching resistance on the surface was removed through SC1 cleaning.

The two types of samples were immersed for 30 min at 81–83 °C in a solution of 2 wt% NaOH, 6.25 wt% Na_2SiO_3, and 12.5 vol% IPA for micro-pyramid texturization [26]. Two types of samples were patterned with the nano–micro-SE structure and micro-pyramid structure. To remove the SiN_X texture barrier film of the patterned nano–micro-SE structure of the samples, these were immersed in BHF for 10 min at room temperature. All the samples were rinsed in flowing DI water, followed by HCl and HF, to remove the metal ions and native oxides on the surface. Finally, the wafers were rinsed again in flowing DI water and dried. The surface morphology of the patterned nano- and micro-pyramid structures was observed using SEM. The emitter layer was formed in a quartz tube using $POCl_3$ at 830 °C for 10 min. The PSG layer was removed and etched on the rear side using the InOxide facility of RENA Technologies GmbH.

For the formation of standardized PERC structures, the front side was passivated using PECVD-SiN_X (single layer with refractive index 2.05), and the rear side was passivated with a stacked layer of atomic layer deposition aluminum oxide (ALD-Al_2O_3) and PECVD-SiN_X. Line-shaped laser contact openings (LCO) were formed on the rear side using a picosecond laser. The Ag front grid metal was printed with an aligned nano-micro-SE patterned with Ag paste. The grid pattern design of the front electrode had 40-µm finger width and 1.312-mm spacing. The rear side was screen-printed on the whole area with a commercially available Al paste, and a furnace firing step completed the PERC cell process. To confirm the Ag printing characteristics of the front electrode formed by each surface structure, the shape of the electrode was observed using an optical microscope (OM), and the line width was measured. The current density-voltage (J–V) measurement of the solar cell fabricated by each surface structure was analyzed under the standard condition of AM 1.5 G (100 mW/cm^2) at 25 °C. The internal quantum efficiency was measured in the wavelength range of 300 to 1100 nm using the QEX7 IPCE(Incident-Photon-to-electron Conversion Efficiency) system (PV Measurements, Inc. in Washington, WA, USA).

3. Results

3.1. Characteristics of Phosphorus Doping According to Nanosurface Morphology

A nanostructured surface with an approximate size of 50 nm was formed through chemical vapor texturing of the cleaned SDR wafers, as shown in Figure 2a,e. Further post-etching of the nanostructured surface for immersion durations of 1, 2, and 3 min changed the sizes of the nanostructures to approximately 100, 200, and 300 nm, respectively. The SEM images in Figure 2a–d show the changed size of the nanostructures formed after vapor-texturing with post-etching. Figure 2e,f shows the cross-section of nano-vapor texturing and micro-pyramid texturing.

In Figure 3, it can be observed that the average emitter sheet resistance for the micro-pyramid structure surface was 98.5 Ω/sq. At this time, for the 50-nm nanostructure surface (shown in Figure 2a,e), the emitter sheet resistance was 61.1 Ω/sq. By changing the nanostructure size to approximately 100, 200, and 300 nm, the emitter sheet resistance was observed to increase to 82.0, 86.4, and 88.8 Ω/sq. As can be observed in Figure 3, owing to the different factors, the sheet resistance for the micro-pyramid structure was 0.64 (nano-vapor texture) for the 50-nm size, 0.81 (post-etching: 1 min) for the 100-nm size, 0.85 (post-etching: 2 min) for the 200-nm size, and 0.88 (post-etching: 3 min) for the 300-nm size. The difference factor index was gradually increased to 1, which was the reference value for the micro-pyramid structure as the increased size of the nanostructure. As a result, the difference factor index for the novel nano-micro-SE structure, being lower than 1, was better.

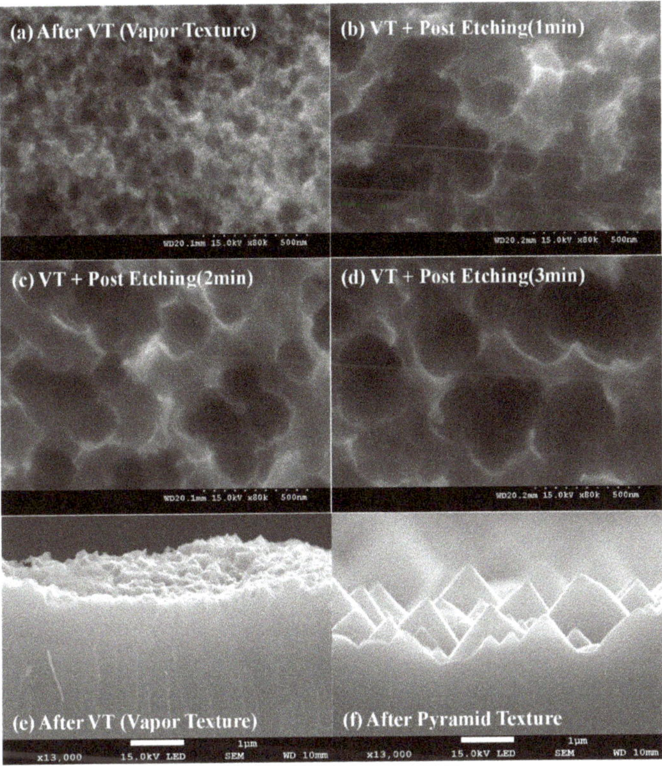

Figure 2. Changes in surface morphology according to nanostructure of vapor texture (**a**) and after post-etching for 1 (**b**), 2 (**c**), and 3 min (**d**); cross-section of nanostructure from the vapored-texture process (**e**); cross-section of micro-pyramid structure from the pyramid textured process (**f**).

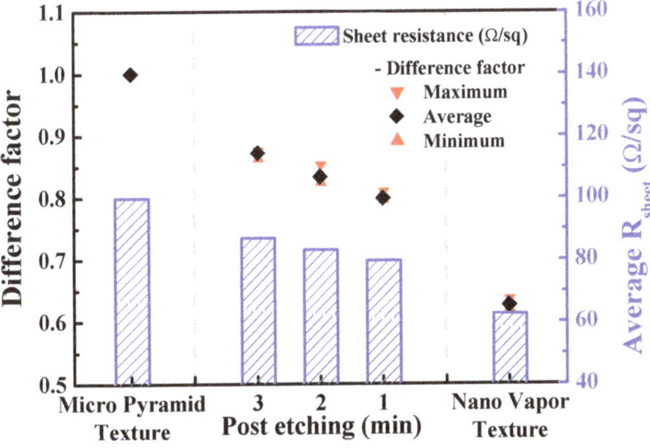

Figure 3. Change in emitter sheet resistance after phosphorus doping process, according to the variation in surface morphology.

3.2. Characteristics of Phosphorus Doping According to Nanosurface Morphology

The samples, with approximately 50-nm size nanostructure on one side and micro-pyramid on the other side, were analyzed with the difference factor index of surface morphology through a doping process about the variation of temperature and time from a high sheet resistance to low sheet resistance. Figure 4 showed the results of the difference factor index of surface morphology by the variation in sheet resistance. For the deep emitter doping conditions, during 30 min at 860 °C, the average sheet resistances obtained for the pyramid-structured and nanostructured surfaces were 40 and 39 Ω/sq, respectively. In this time, the difference factor index for the 50 nm nanostructured- and micro-pyramid-structured surfaces was 0.97, which was attributed to almost equal sheet resistance characteristics for both surfaces. Figure 4 shows the results, and it is observed that as the sheet resistance increased by the doping process condition from a low sheet resistance, the difference factor index became smaller and the SE characteristic was increased. The best difference factor index region of the SE characteristic was when the doping condition with the sheet resistance for the pyramid structure surface was approximately 100 to 150 Ω/sq, and for the nanostructured surface, the range was about 74 to 122 Ω/sq with the difference factor converged to the level of 0.76. This result was compared with the difference factor index of 0.64 shown in Figure 3. The nano-vapor texture was predicted to result from some large-sized nanostructure. Further, sheet resistance of more than 150 Ω/sq on the pyramid-structured surface was obtained by the doping process condition, and the difference factor index increased for a sheet resistance up to 360 Ω/sq. The experimental results have shown a sheet resistance of approximately 360 Ω/sq for the shallow emitter, and limited change occurred in the sheet resistance according to the surface morphology, whereas the deep emitter had a sheet resistance of 40 Ω/sq.

The variation in sheet resistance for the doping condition between the nanostructures and the micro-pyramid structure (as shown in Figure 3) indicated a reduced to maximum sheet resistance in the range of about 100 to 150 Ω/sq by the size of the nanostructure. It can be seen that the SE characteristic was maximized in this doping region. Shown in Figure 4, the sheet resistance of the "A" sample as the shallow emitter and the "B" sample as the general emitter were analyzed by SIMS to investigate the factors of change in sheet resistance of the nano- and micro-surface structure according to the doping condition.

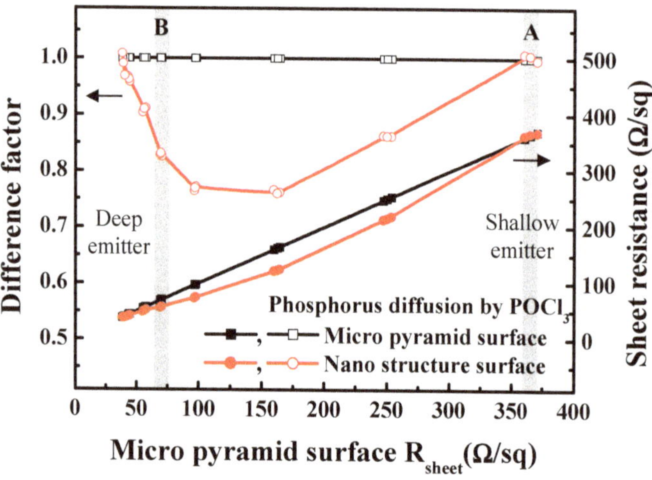

Figure 4. Change in emitter sheet resistance of micro-pyramid structure and nanostructure, according to the doping process condition.

The SIMS result in Figure 5 indicated that the analysis of the surface morphology of the sample can be measured by the change in silicon (Si) intensity and phosphorus (P) concentration according to the sputter time. On the surface of the nanostructures shown in Figure 5a,c, the signal of the initial Si was low, and it was detected as 3×10^3 cps level. After that, it increased rapidly and then stabilized at 6×10^3 cps level over 300 s, and an intensity of 7×10^3 cps level was detected over 550 s. On the surface of the micro-pyramid structure shown in Figure 5b,d, the low signal of the initial Si was higher, and it was detected at the level of 6×10^3 and 4×10^3 cps compared with the nanosurface. It can be observed that it is relatively easier to detect Si than the nanostructure surface. In other words, the doped silicon surface of the nanostructure morphology was coated with many dopant materials, more than the micro-pyramid surface.

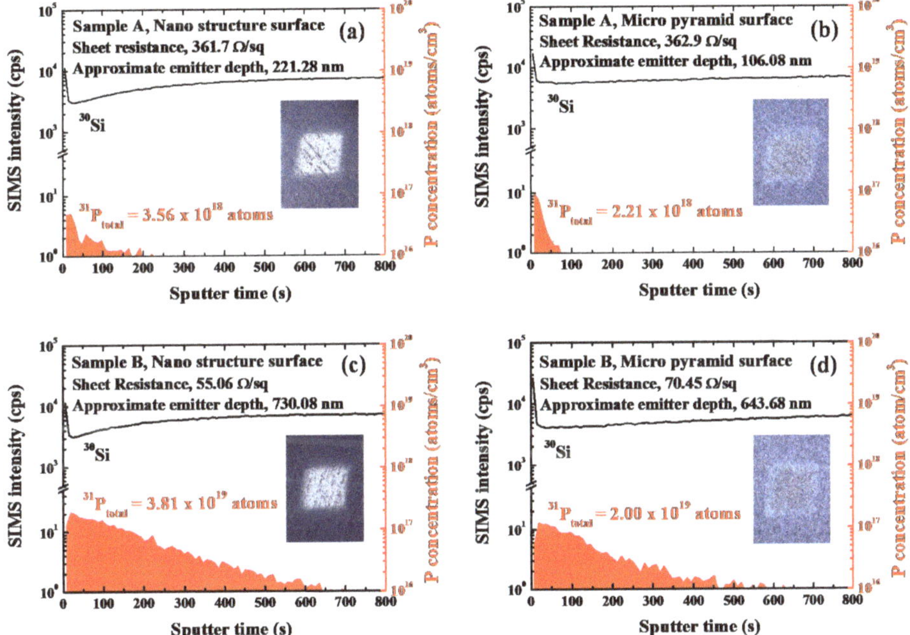

Figure 5. Secondary ion mass spectroscope (SIMS) analysis of surface morphology according to the doping process condition, (**a**) nanostructure surface in shallow-doping condition A, (**b**) micro-pyramid structure surface in shallow-doping condition A, (**c**) nanostructure surface in traditional-doping condition B, (**d**) micro-pyramid structure surface in traditional-doping condition B.

The results of the analysis of the P concentration by depth to SIMS analysis did not show accurate values due to the shape of the surface structure [27]. However, the results of that detected by SIMS are valid in the comparative analysis of the total amount of P in the pyramid structure and the nanostructure. In Figure 5a, for the "A" sample of nanostructure morphology, the total amount of P detection concentration is 3.56×10^{18} atoms/cm^3 at a sheet resistance of 361.7 Ω/sq. Regarding the micro-pyramid morphology of the "A" sample, the total amount of P detection concentration was 2.21×10^{18} atoms/cm^3 and the sheet resistance was 362.9 Ω/sq, as shown in Figure 5b. In the similar sheet resistance, approximately 360 Ω/sq by shallow doping was the highest total amount of detected concentration of P for the nanostructure morphology compared with the micro-pyramid structure. This result is due to the fact that many dopant materials were inactive on the large surface area of the nanostructure surface by the shallow doping process [28].

In Figure 5c, for the "B" sample with the SE characteristic of nanostructure morphology, the total amount of P detection concentration was 3.81×10^{19} atoms/cm^3 at a sheet resistance of 55.06 Ω/sq. For the micro-pyramid surface morphology of "B" samples, the total amount of P detection concentration was 2.00×10^{19} atoms/cm^3 and sheet resistance 70.45 Ω/sq, as shown in Figure 5d. Even under the doping conditions of "B" samples with the SE characteristic, the total amount of P doped with the surface characteristics of the nanostructure was higher than that of the pyramid structure. In deep doping conditions, the emitter layer became deeper, and the emitter from the deeper part of the nanostructure was fused [29,30]. As a result, for the deeper emitter doping conditions, it had a similar sheet resistance of approximately 40 Ω/sq and a difference factor index of approximately 1 for the nanostructured surface and the micro-textured surface, as shown in Figure 4.

3.3. *Mechanism of SE Formation by Nano–Micro-Morphology*

The results of sheet resistance analysis in Figures 3 and 4 clearly shows that the best SE structure had a smaller size of the nanostructure, and the doping process condition for the sheet resistance region on the pyramid structure was 100–150 Ω/sq.

Figure 6 shows the SE mechanism in which nanostructures and micro-pyramid structures were formed through the doping process. The local nanostructures for SEs were formed at the texture processing stage with pyramidal structures. The formed textured surface structure had two surface structures simultaneously, as shown in Figure 6a. To produce emitter layer formation, the phosphorus silicate glass (PSG) layer was formed on the surface of the silicon by a pre-deposition process of $POCl_3$ doping in a quartz tube furnace, as shown in Figure 6b. The report of Catherine et al. analyzes in detail the thickness of the formed PSG layer and the depth of the emitter formed in the peaks, valleys, and flank of the pyramid structure during the doping process [31]. It can be seen from the literature that the PSG layer in the valley position "A" is formed relatively thick in the pyramid structure compared to the peak and flank as shown in Figure 6b. In the nanostructure, it can be understood that the thick PSG layer formed more than the pyramid structure thick PSG layer because the valley position "A" had more frequency than the pyramid structure with respect to the same area. As a result, the doped silicon surface of the nanostructure morphology was coated with many dopant materials more than the surface of the micro-pyramid. Dastgheib-Shirazi et al. reported an analysis of the depth variation of the thermally diffused emitter according to the thickness of the PSG layer formed in the doping process [32]. It can be seen from the literature that the thick PSG layer formed a higher surface concentration of phosphorus, forming a deeper emitter layer through the thermal diffusion of the doping process, as shown in Figure 6c. Moreover, it can be seen from the literature that deeper emitters for the micro-pyramid structure were formed in the peak "D" than in the valleys "B" and flanks "C" [31]. In the nanostructure region, it can be understood that the sheet resistance was lower than that in the micro-pyramid structure formed by a higher frequency structure valley "A" of a thicker PSG layer and the deep diffused layer of peak "D" through the emitter doping process. In conclusion, the surface morphology of nanostructure and micro-pyramid structure was through the conventional thermal diffusion emitter doping process becoming the novel SE structure as shown in Figure 7.

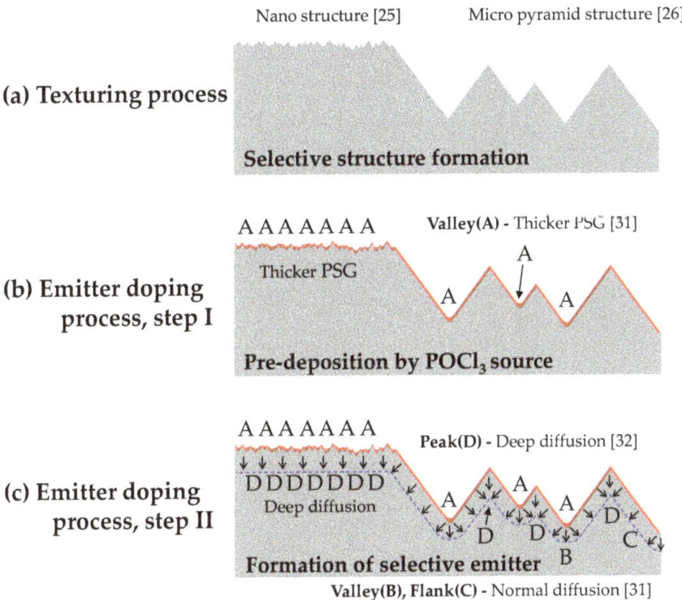

Figure 6. Formation mechanism of the novel selective emitter (SE) according to surface morphology, (**a**) selective structure formation on texturing process [25,26], (**b**) phosphorus pre-deposition on doping process [31], (**c**) phosphorus diffusion on doping process [31,32].

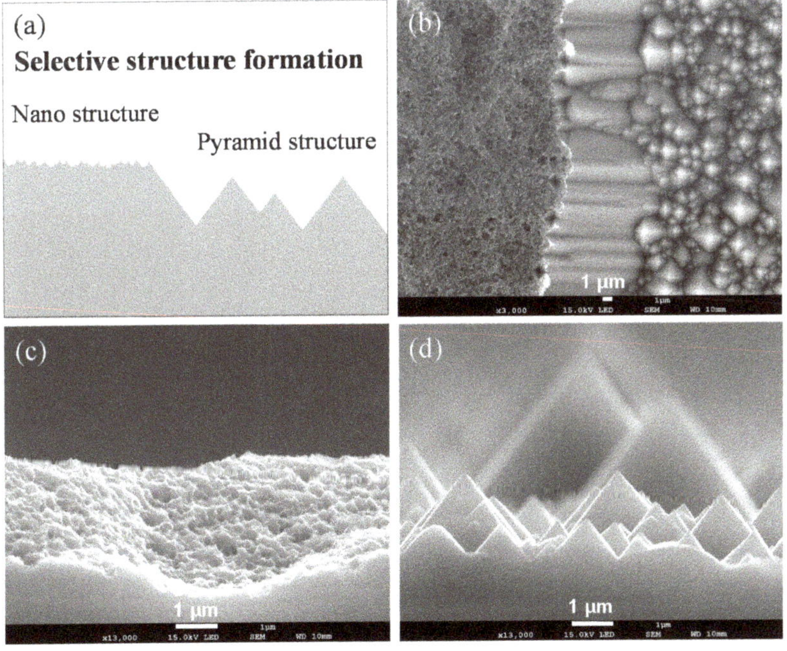

Figure 7. SEM image of the novel SE surface using a nanostructure and a micro-pyramid structure, (**a**) schematic diagram of the novel SE structure, (**b**) top view of the novel SE surface in 3 K magnification,

(**c**) cross-section view of the nanostructure side in 13 K magnification, (**d**) cross-section view of the micro-pyramid structure side in 13 K magnification. Included in all is scale bar of 1 μm.

3.4. Fabrication of Novel SE Solar Cell.

Figure 7 shows the schematic diagram Figure 7a of the SE using a nano-micro-pyramid structure and the SEM analysis of the fabricated samples. Figure 7b shows the surface of the novel SE structure sample analyzed through SEM at 3 K magnification on which the nanostructure and micro-pyramid structure were formed together. Figure 7c,d show the enlarged SEM result of the cross-section of the nanostructure and micro-pyramid structure at 13 K magnification. In the nanostructure of Figure 7c, as described in the mechanism of SE formation by the surface morphology of Figure 6, the valley and the peak observed had a higher frequency compared with the micro-pyramid structure in Figure 7d.

Table 1 and Figure 8 show the best J–V characteristics of the solar cells fabricated with nanostructure cells (NSC), micro-pyramid structure cells (MPC), and the novel SE structure cells (SEC), respectively. The J–V curve results from the NSC emitter had a sheet resistance of 62 Ω/sq, short-circuit current density (J_{SC}) of 39.81 mA/cm^2, open-circuit voltage (V_{OC}) of 638 mV, fill factor (FF) of 79.82%, and conversion efficiency of 20.27%. The J–V curve results from the MPC emitter had sheet resistance of 82 Ω/sq, short-circuit current density (J_{SC}) of 40.35 mA/cm^2, open-circuit voltage (V_{OC}) of 647 mV, fill factor (FF) of 79.00%, and conversion efficiency of 20.61%. The NSC showed a short-circuit current density (J_{SC}) of 0.54 mA/cm^2 and an open-circuit voltage (V_{OC}) of 9 mV, which was degraded compared to that of the MPC. As shown in this result, it was difficult to emitter passivate the nanostructure surface compared to the micro-pyramid surface [29]. However, the NSC showed 0.82% improvement in the fill factor (FF) compared to the MPC. The improved fill factor (FF) in the NSC can be seen in the literature as an improved series resistance by electrode contact with the sheet resistance of the front emitter [33]. Despite improvements in the fill factor (FF) of the NSC, the degradation of efficiency characteristics due to short circuit current density (J_{SC}) and open-circuit voltage (V_{OC}) was due to the low sheet resistance characteristics and higher recombination characteristics [29].

Table 1. Best and average solar cell characteristics according to surface structure with a comparison of sheet resistance.

Light J–V Analysis	**Nanostructure Cell (NSC)**		**Micro-Pyramid Cell (MPC)**		**SE Cell (SEC) Low/High**	
Sheet Resistance of Emitter (Ω/sq)	62		82		62/82	
Cell remark (9 samples fabricated)	Best	Average	Best	Average	Best	Average
J_{SC} (mA/cm^2)	39.81	39.07	40.35	40.24	40.79	40.63
V_{OC} (V)	0.638	0.637	0.647	0.643	0.643	0.642
Fill Factor (%)	79.82	80.18	79.00	78.61	79.60	79.29
Efficiency (%)	20.27	19.94	20.61	20.35	20.88	20.67

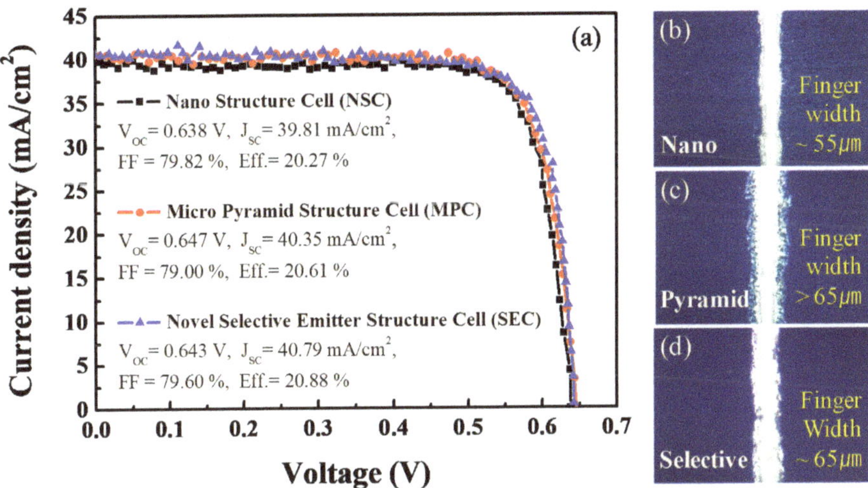

Figure 8. Light density–voltage (J–V) characteristic analysis of crystalline silicon (c-Si) solar cells according to surface morphology and printed front Ag electrode shape, (**a**) result of light J–V, (**b**) optical microscope (OM) image of printed front Ag electrode shape on nanostructured solar cell surface, (**c**) OM image of printed front Ag electrode shape on micro-pyramid-structured solar cell surface, (**d**) OM image of printed front Ag electrode shape on the novel SE-structured solar cell surface.

The J–V curve results for the novel SE structure cell (SEC) using nanostructure and micro-pyramid structure had a current density (J_{SC}) of 40.79 mA/cm^2, open-circuit voltage (V_{OC}) of 643 mV, fill factor (FF) of 79.60%, and conversion efficiency of 20.88%. The improvement in the SEC efficiency of the selective surface structure emitter was analyzed by an FF increase (0.6%) and a J_{SC} increase (0.44 mA/cm^2) with respect to those of the MPC. The improvement in the fill factor (FF) was also observed in the internal quantum efficiency (IQE) results of Figure 9 due to the low contact resistance of the nanostructure electrode contacts.

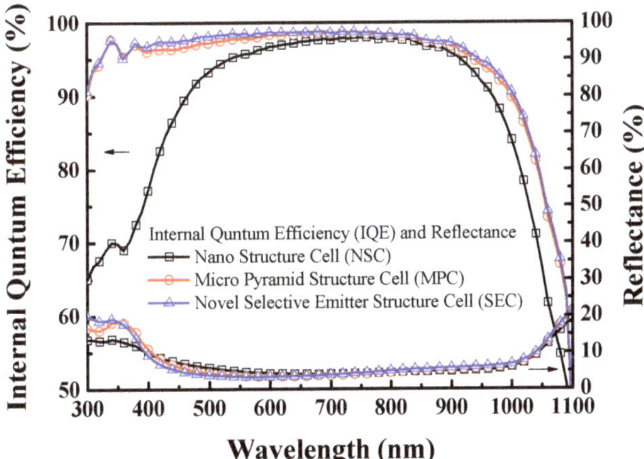

Figure 9. Result of reflectance and internal quantum efficiency (IQE) of c-Si solar cells according to surface morphology.

In the 300 to 400 nm region observed as the recombination characteristics of the surface emitter, similar analytical results as the micro-pyramid structure were shown, but improved IQE of average

0.56% and maximum 1.1% in the wavelength range of 400 to 1100 nm were obtained. The improvement in the IQE shows the characteristic of decreasing the series resistance, which also affects the improvement in the fill factor (FF) [34]. The improvement in the short circuit current density (J_{SC}) is confirmed by the improvement in the shading loss of the screen-printing electrode according to the surface structure of Figure 8b–d. The screen-printed electrode lines on the nanostructure surface showed a clean edge electrode shape of about 55 μm shown in Figure 8b. In the micro-pyramid structure of Figure 8c, the screen-printed electrode line width was approximately 65 μm or more by electrode spreading. It can be seen from the literature that the short-circuit current density was lowered due to the shading loss by the electrode spreading caused by the action of the Ag particle size of the screen-printing electrode and the micro-pyramid structure size [35]. As shown in Figure 8d, the printed electrode line width of the selective surface structure was approximately 65 μm and a clean edge electrode shape was observed. The shape of the clean edge electrode by the selective surface structure minimized the shading loss and could obtain improved characteristics of short circuit current density (J_{SC}) compared to a micro-pyramid structure.

In this study, we fabricated a new SE solar cell using the final nanostructure and micro-pyramid structure. This was performed by controlling the difference in sheet resistance in the doping process together with the pyramid structure formed in the texturing process using the nanostructure formed before the texturing process. As the size of the nanostructure became smaller, the emitter sheet resistance of the doped sample became smaller. The difference in sheet resistance between the nanostructure and the pyramid structure was found to be the largest in the range of the emitter doping condition of 100 to 150 Ω/sq at the micro-pyramid structure. These experimental facts have been confirmed by the SIMS analysis and the literature, which show that the nanostructure had more frequent valleys and peaks than that of the micro-pyramid structure. As a result of the characteristics of the solar cell having the novel SE using the nanostructure, the efficiency of 20.88% was improved by about 0.27% compared to the conventional pyramid structure with the efficiency of 20.61%. This improvement was obtained by improving the fill factor (FF) through the contact of the electrode with the low sheet resistance surface on the nanostructure. It also obtained an improvement in the short circuit current density (J_{SC}) by minimizing the shadow loss of the electrode printed on the nanostructure surface.

4. Conclusions

In this study, a novel SE technique that uses selective nanosurface morphologies for a heavy emitter area was proposed before the pyramid texturing process [3–24]. The SE technology, which used different surface morphologies, exhibited different sheet resistances despite the nano/micro-structure regions being doped with a single doping process. It was also possible to control the difference in sheet resistance during the doping process by changing the temperature, pre-deposition time, and drive-in time. The result of the uniformly doped sheet resistance on the standard micro-pyramidal surface was about 82 Ω/sq, and the doping results of the selectively formed nano/micro-surface were 62 and 82 Ω/sq, respectively. As a result, doped-on selectively formed nano/micro-surfaces SE cells showed a J_{SC} increase (0.44 mA/cm^2) and an FF increase (0.6%) compared to the homogeneously doped cells on the micro-pyramid surface, which resulted in about 0.27% enhanced conversion efficiency.

The novel SE technology that uses nano/micro-structures with similar form to the selective epitaxial layer growth SE technology blocked the risk of solar cell manufacturing processes such as contamination, surface damage, reflectance loss, and over-alignment loss. At each different height location, the nanostructure and the micro-pyramid structure were formed on the silicon surface. Therefore, alignment loss could be minimized by easily aligning the front printed electrode to the high-level location region on the nanosurface.

Author Contributions: Conceptualization, M.J.; methodology, M.J.; validation, Y.H.C., Y.K., and E.-C.C.; formal analysis, M.J., J.P., and Y.H.C.; investigation, M.J.; data curation, M.J., J.P., and Y.H.C.; writing—original draft preparation, M.J. and Y.K.; writing—review and editing, M.J., Y.K., and E.-C.C.; supervision, D.L., E.-C.C., and J.Y.;

project administration, E.-C.C.; funding acquisition, D.L, E.-C.C., and J.Y. All authors have read and agreed to the published version of the manuscript.

Funding: This work was supported by the Korea Institute of Energy Technology Evaluation and Planning (KETEP) and the Ministry of Trade, Industry and Energy (MOTIE) of the Republic of Korea (No. 20203030010310 and No. 20173010013740).

Conflicts of Interest: The authors declare no conflict of interest.

References

1. Green, M.A. The path to 25% silicon solar cell efficiency: History of silicon cell evolution. *Prog. Photovolt. Res. Appl.* **2009**, *17*, 183–189. [CrossRef]
2. Swanson, R.M. A vision for crystalline silicon photovoltaics. *Prog. Photovolt. Res. Appl.* **2006**, *14*, 443–453. [CrossRef]
3. ITRPV 2020. *International Technology Roadmap for Photovoltaic (ITRPV) Eleventh Edition—2019 Results*; ITRPV: Frankfurt, Germany, 2020.
4. Gassenbauer, Y.; Ramspeck, K.; Bethmann, B.; Dressler, K.; Moschner, J.D.; Fiedler, M.; Brouwer, E.; Drößler, R.; Lenck, N.; Heyer, F.; et al. Rear-Surface Passivation Technology for Crystalline Silicon Solar Sells: A Versatile Process for Mass Production. *IEEE J. Photovolt.* **2013**, *3*, 125–130. [CrossRef]
5. Green, M.A. The passivated emitter and rear cell (PERC): From conception to mass production. *Sol. Energy Mater. Sol. Cells* **2015**, *143*, 190–197. [CrossRef]
6. Chunduri, S.K.; Schmela, M. High Efficiency Cell Technologies 2019 From PERC to Passicated Contacts and HJT. *Taiyang News*, 2019.
7. Min, B.; Wagner, H.; Müller, M.; Neuhaus, H.; Brendel, R.; Altermatt, P.P. Incremental efficiency improvements of mass-produced PERC cells up to 24%, predicted solely with continuous development of existing technologies and wafer materials. In Proceedings of the 31st European Photovoltaic Solar Energy Conference and Exhibition, Hamburg, Germany, 14–18 September 2015; pp. 473–476.
8. Hahn, G. Status of selective emitter technology. In Proceedings of the 25th European Photovoltaic Solar Energy Conference and Exhibition and 5th World Conference on photovoltaic Energy Conversion, Munich, Germany, 6–10 September 2010; pp. 1091–1096.
9. Rahman, M.Z. Status of selective emitters for p-type c-Si solar cells. *Opt. Photonics J.* **2012**, *2*, 129–134. [CrossRef]
10. Lv, Y.; Zhuang, Y.F.; Wang, W.J.; Wei, W.W.; Sheng, J.; Zhang, S.; Shen, W.Z. Towards high-efficiency industrial p-type mono-like Si PERC solar cells. *Sol. Energy Mater. Sol. Cells* **2020**, *204*, 110202. [CrossRef]
11. Tonini, D.; Borrosso, C.; Cellere, G.; Furin, V.; Galiazzo, M.; Kumar, P.; Tanner, D.; Voltan, A. Efficiency gain in c-Si cells through selective emitter and double printing. *Energy Procedia* **2011**, *8*, 598–606. [CrossRef]
12. Horzel, J.; Szlufcik, J.; Nijs, J.; Mertens, R. A simple processing sequence for selective emitters [Si solar cells]. In Proceedings of the Conference Record of the Twenty Sixth IEEE Photovoltaic Specialists Conference, Anaheim, CA, USA, 29 September–3 October 1997; pp. 139–142.
13. Antoniadis, H. Silicon ink high efficiency solar cells. In Proceedings of the 34th IEEE Photovoltaic Specialists Conference, Philadelphia, PA, USA, 7–12 June 2009; pp. 000650–000654.
14. Zhong, S.; Shen, W.; Liu, F.; Li, X. Mass production of high efficiency selective emitter crystalline silicon solar cells employing phosphorus ink technology. *Sol. Energy Mater. Sol. Cells* **2013**, *117*, 483–488. [CrossRef]
15. Lauermann, T.; Book, F.; Dastgheib-Shirazi, A.; Hahn, G.; Haverkamp, H.; Bleidiessel, R.; Fleuster, M. The optimal choice of the doping levels in an inline selective emitter design for screen printed multicrystalline silicon solar cells. In Proceedings of the 24th European Photovoltaic Solar Energy Conference and Exhibition, Hamburg, Germany, 21–24 September 2009; pp. 1795–1797.
16. Basu, P.K.; Cunnusamy, J.; Sarangi, D.; Boreland, M.B. Novel selective emitter process using non-acidic etch-back for inline-diffused silicon wafer solar cells. *Renew. Energy* **2014**, *66*, 69–77. [CrossRef]
17. Book, F.; Raabe, B.; Hahn, G. Two diffusion step selective emitter: Comparison of mask opening by laser or etching paste. In Proceedings of the 23rd European Photovoltaic Solar Energy Conference and Exhibition, Valencia, Spain, 1–5 September 2008; pp. 1546–1549.

18. Engelhardt, J.; Kromer, H.; Hahn, G.; Terheiden, B. Laser doping from as-deposited CVD layers for high-efficiency crystalline silicon solar cells. In Proceedings of the Silicon PV 2019, the 9th International Conference on Crystalline Silicon Photovoltaics, Leuven, Belgium, 8–10 April 2019; Volume 2147, p. 070002.
19. Kim, M.; Kim, D.; Kim, D.; Kang, Y. Influence of laser damage on the performance of selective emitter solar cell fabricated using laser doping process. *Sol. Energy Mater. Sol. Cells* **2015**, *132*, 215–220. [CrossRef]
20. Raabe, B.; Haverkamp, H.; Book, F.; Dastgheib-Shirazi, A.; Moll, R.; Hahn, G. Monocrystalline silicon: Future cell concepts. In Proceedings of the 22nd European Photovoltaic Solar Energy Conference and Exhibition, Milano, Italy, 3–7 September 2007; pp. 1024–1029.
21. Pal, B.; Ray, S.; Gangopadhyay, U.; Ray, P.P. Novel technique for fabrication of n-type crystalline silicon selective emitter for solar cell processing. *Mater. Res. Express* **2019**, *6*, 075523. [CrossRef]
22. Low, R.; Gupta, A.; Bateman, N.; Ramappa, D.; Sullivan, P.; Skinner, W.; Mullin, J.; Peters, S.; Weiss-Wallrath, H. High efficiency selective emitter enabled through patterned ion implantation. In Proceedings of the 35th IEEE Photovoltaic Specialists Conference, Honolulu, HI, USA, 20–25 June 2010; pp. 001440–001445.
23. Dubé, C.E.; Tsefrekas, B.; Buzby, D.; Tavares, R.; Zhang, W.; Gupta, A.; Low, R.J.; Skinner, W.; Mullin, J. High efficiency selective emitter cells using patterned ion implantation. *Energy Procedia* **2011**, *8*, 706–711. [CrossRef]
24. Payo, M.R.; Li, Y.; Russell, R.; Singh, S.; Filipek, I.K.; Duerinckx, F.; Szlufcik, J.; Poortmans, J. Efficiency gain in plated bifacial n-type PERT cells by means of a selective emitter approach using selective epitaxy. *Sol. Energy Mater. Sol. Cells* **2020**, *204*, 110173. [CrossRef]
25. Ju, M.; Balaji, N.; Lee, Y.; Park, C.; Song, K.; Choi, J.; Yi, J. Novel vapor texturing method for EFG silicon solar cell applications. *Sol. Energy Mater. Sol. Cells* **2012**, *107*, 366–372. [CrossRef]
26. Ju, M.; Balaji, N.; Park, C.; Nguyen, H.T.T.; Cui, J.; Oh, D.; Jeon, M.; Kang, J.; Shim, G.; Yi, J. The effect of small pyramid texturing on the enhanced passivation and efficiency of single c-Si solar cells. *RSC Adv.* **2016**, *6*, 49831–49838. [CrossRef]
27. Komatsu, Y.; Harata, D.; Schuring, E.W.; Vlooswijk, A.H.G.; Katori, S.; Fujita, S.; Venema, P.R.; Cesar, I. Calibration of electrochemical capacitance-voltage method on pyramid texture surface using scanning electron microscopy. *Energy Procedia* **2013**, *38*, 94–100. [CrossRef]
28. Li, H.; Ma, F.; Hameiri, Z.; Wenham, S.; Abbott, M. On elimination of inactive phosphorus in industrial $POCl_3$ diffused emitters for high efficiency silicon solar cells. *Sol. Energy Mater. Sol. Cells* **2017**, *171*, 213–221. [CrossRef]
29. Kafle, B.; Schön, J.; Fleischmann, C.; Werner, S.; Wolf, A.; Clochard, L.; Duffy, E.; Hofmann, M.; Rentsch, J. On the emitter formation in nanotextured silicon solar cells to achieve improved electrical performances. *Sol. Energy Mater. Sol. Cells* **2016**, *152*, 94–102. [CrossRef]
30. Es, F.; Baytemir, G.; Kulakci, M.; Turan, R. Metal-assisted nano-textured solar cells with SiO_2/Si_3N_4 passivation. *Sol. Energy Mater. Sol. Cells* **2017**, *160*, 269–274. [CrossRef]
31. Voyer, C.; Buettner, T.; Bock, R.; Biro, D.; Preu, R. Microscopic homogeneity of emitters formed on textured silicon using in-line diffusion and phosphoric acid as the dopant source. *Sol. Energy Mater. Sol. Cells* **2009**, *93*, 932–935. [CrossRef]
32. Dastgheib-Shirazi, A.; Steyer, M.; Micard, G.; Wagner, H.; Altermatt, P.P.; Hahn, G. Relationships between diffusion parameters and phosphorus precipitation during the $POCl_3$ diffusion process. *Energy Procedia* **2013**, *38*, 254–262. [CrossRef]
33. Shanmugam, V.; Cunnusamy, J.; Khanna, A.; Basu, P.K.; Zhang, Y.; Chen, C.; Stassen, A.F.; Boreland, M.B.; Mueller, T.; Hoex, B.; et al. Electrical and microstructural analysis of contact formation on lightly doped phosphorus emitters using thick-film Ag screen printing pastes. *IEEE J. Photovolt.* **2013**, *4*, 168–174. [CrossRef]
34. Yang, Y.; Seyedmohammadi, S.; Kumar, U.; Gnizak, D.; Graddy, E.; Shaikh, A. Screen printable silver paste for silicon solar cells with high sheet resistance emitters. *Energy Procedia* **2011**, *8*, 607–613. [CrossRef]
35. Ju, M.; Mallem, K.; Dutta, S.; Balaji, N.; Oh, D.; Cho, E.; Cho, Y.H.; Kim, Y.; Yi, J. Influence of small size pyramid texturing on contact shading loss and performance analysis of Ag-screen printed mono crystalline silicon solar cells. *Mater. Sci. Semicond. Process.* **2018**, *85*, 68–75. [CrossRef]

Article

Complex Investigation of High Efficiency and Reliable Heterojunction Solar Cell Based on an Improved Cu_2O Absorber Layer

Laurentiu Fara [1,2,*], Irinela Chilibon [3], Ørnulf Nordseth [4], Dan Craciunescu [1], Dan Savastru [3], Cristina Vasiliu [3], Laurentiu Baschir [3], Silvian Fara [1], Raj Kumar [5], Edouard Monakhov [5] and James P. Connolly [6]

1 Department of Physics, Faculty of Applied Sciences, Polytechnic University of Bucharest, 060042 Bucharest, Romania; dan.craciunescu@sdettib.pub.ro (D.C.); silvian.fara@gmail.com (S.F.)
2 Academy of Romanian Scientists, 050091 Bucharest, Romania
3 National Institute of Research and Development for Optoelectronics (INOE-2000), 077125 Bucharest, Romania; chilib@inoe.ro (I.C.); dsavas@inoe.ro (D.S.); icvasiliu@inoe.ro (C.V.); baschirlaurentiu@inoe.ro (L.B.)
4 Institute for Energy Technology (IFE), P.O. Box 40, NO-2027 Kjeller, Norway; Ornulf.Nordseth@ife.no
5 Department of Physics/Center for Materials Science and Nanotechnology (SMN), University of Oslo, P.O. Box 1048 Blindern, N-0316 Oslo, Norway; raj.kumar@smn.uio.no (R.K.); eduard.monakhov@fys.uio.no (E.M.)
6 GeePs (Group of electrical engineering—Paris), UMR CNRS 8507, CentraleSupélec, Univ. Paris-Sud, Université Paris-Saclay, Sorbonne Université, 11 rue Joliot-Curie, 91192 Plateau de Moulon, Gif-sur-Yvette CEDEX, France; jpgconnolly@gmail.com
* Correspondence: lfara@renerg.pub.ro

Received: 30 July 2020; Accepted: 31 August 2020; Published: 8 September 2020

Abstract: This study is aimed at increasing the performance and reliability of silicon-based heterojunction solar cells with advanced methods. This is achieved by a numerical electro-optical modeling and reliability analysis for such solar cells correlated with experimental analysis of the Cu_2O absorber layer. It yields the optimization of a silicon tandem heterojunction solar cell based on a ZnO/Cu_2O subcell and a c-Si bottom subcell using electro-optical numerical modeling. The buffer layer affinity and mobility together with a low conduction band offset for the heterojunction are discussed, as well as spectral properties of the device model. Experimental research of N-doped Cu_2O thin films was dedicated to two main activities: (1) fabrication of specific samples by DC magnetron sputtering and (2) detailed characterization of the analyzed samples. This last investigation was based on advanced techniques: morphological (scanning electron microscopy—SEM and atomic force microscopy—AFM), structural (X-ray diffraction—XRD), and optical (spectroscopic ellipsometry—SE and Fourier-transform infrared spectroscopy—FTIR). This approach qualified the heterojunction solar cell based on cuprous oxide with nitrogen as an attractive candidate for high-performance solar devices. A reliability analysis based on Weibull statistical distribution establishes the degradation degree and failure rate of the studied solar cells under stress and under standard conditions.

Keywords: silicon tandem heterojunction solar cell; N-doped Cu_2O absorber layer; Al:ZnO (AZO); numerical electro-optical modeling; scanning electron microscopy (SEM); atomic force microscopy (AFM); X-ray diffraction (XRD); spectroscopic ellipsometry (SE); Fourier-transform infrared (FTIR) spectroscopy; degradation degree; failure rate

---

1. Introduction

Cuprous oxide has attracted much attention for thin film photovoltaic applications due to its high absorption coefficient in the visible region [1]. It is also a nontoxic, low cost material, which is abundant in the Earth's crust. Cu_2O crystallizes in a cubic structure with a lattice constant of 4.269 Å, where the copper (Cu) and oxygen (O) atoms are arranged in face-centered cubic (fcc) and body-centered cubic (bcc) sub-lattices, respectively. Cu_2O is a natural p-type semiconductor with a direct band gap of about 2.1 eV, with negatively charged Cu vacancies and O interstitials [2].

Concerning fabrication, cuprous oxide thin films have been prepared by a wide range of synthesis techniques including reactive sputtering, sol–gel technique, plasma evaporation, thermal oxidation, chemical vapor deposition, anodic oxidation, reactive sputtering, electrodeposition, and so on [3]. Wang et al. have demonstrated that the incorporation of nitrogen in thin films can significantly decrease the resistivity [4]. Nitrogen doping in Cu_2O has been shown to provide an effective approach to reducing the resistivity/enhancing the hole concentration, as the reported carrier concentrations in N-doped Cu_2O thin films are of the order of 10^{16}–10^{18} cm^{-3}.

Sberna et al. [5] analyzed the effects of nitrogen doping, introduced by ion implantation, on the parameters of Cu_2O films. They show that the optimization of structural, optical, and electrical properties of Cu_2O thin films synthesized by non-reactive magnetron sputtering can be achieved by post-deposition thermal annealing in oxygen atmosphere.

Foremost among the challenges in metal oxide solar cells development are the improvement of the performance and reliability of these devices. A potential candidate for high-efficiency metal oxide solar cells, such as the Cu_2O/ZnO/c-Si tandem heterojunction solar cell, could theoretically reach a conversion efficiency above 30% [6]. The key to achieving higher efficiencies for solar cells made from these materials is to tune the optical and electrical properties of absorber material in the solar cell, such as cuprous oxide (Cu_2O).

The conversion efficiency for the current ZnO/Cu_2O heterojunction solar cells is around 8%, while the potential is 14% conversion efficiency. Factors like low power conversion efficiency, high defect density, and large conduction band offset for the metal oxide heterojunction imply that there is room for research to further optimize the device. Probe-scanning instruments can provide valuable information towards the characterization and optimization of solar cell layers. However, interface defect states pose a challenge in reaching efficiencies close to the theoretical maximum. Currently, the conversion efficiency for cuprous oxide heterojunctions has the potential to reach 14% as reported by Takiguchi et al. [7] and a theoretical maximum efficiency of around 19%.

The present study is based on numerical electro-optical modeling and reliability analysis and is dedicated to increasing the performance and reliability of heterojunction solar cells [8–10]; it is correlated with the experimental investigation (morphological, structural, and optical) of an improved Cu_2O absorber layer. Its morphological and structural characterization is based on SEM, AFM, and XRD analyses [11]. The optical characterization of the Cu_2O layer uses FTIR spectroscopy and SE analyses [12]. Furthermore, a statistical reliability study is carried out, analyzing the lifetime, degradation degree, and failure rate of the studied solar cells.

2. Numerical Modeling of Cu_2O Heterojunction Solar Cell

2.1. Diagram of the Solar Cell Simulation Model

A schematic representation of the Cu_2O heterojunction solar cell simulation model is shown in Figure 1. The device structure includes a quartz glass superstrate, an aluminum-doped ZnO (AZO) n+ emitter layer, a Cu_2O p-type absorber layer, and a bottom Cu_2O:N layer.

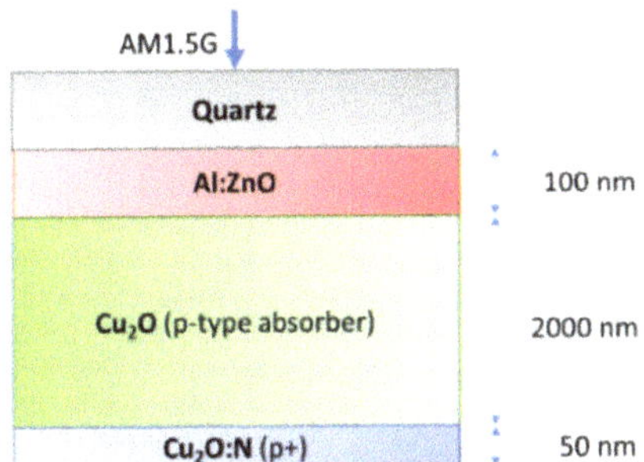

Figure 1. Simplified diagram of the simulated metal oxide solar cell.

2.2. Electrical Modeling and Simulation Results for the Two Subcells of the Tandem Heterojunction Solar Cell

Numerical modeling of the metal oxide heterojunction solar cell was developed based on three simulation software packages, namely Silvaco Atlas, MATLAB, and Quokka 2. The tandem heterojunction solar cell was constituted from two subcells: the metal oxide (Cu_2O) top subcell and c-Si bottom subcell. The electrical properties of the absorber (Cu_2O) and buffer layer (ZnO) from the structure of the top subcell were analyzed using Silvaco Atlas software. The implementation of defects, as well as the study of two efficient materials (ZnO and Ga_2O_3) for the buffer layer were investigated [13–17] and several analyses were considered using the Takiguchi model [7]. A transition interface defect layer (IDL) was introduced as well.

Several comparisons using the Silvaco numerical model were discussed. One of the most interesting buffer materials is ZnO generally, and in particular, the AZO layer (Al:ZnO). This material is very suitable from the point of view of the conduction bandgap offset in regards to Cu_2O. The dependence of the AZO thickness on top cell efficiency is plotted in the Figure 2; the optimum value for the AZO thickness looks to be 0.25 µm. Further analysis examined (1) the dependence of IDL thickness on top cell efficiency (Figure 3), (2) the impact of buffer electron mobility on subcell fill factor (FF) and efficiency (η) (Figure 4), and (3) the impact of buffer and IDL affinity on subcell efficiency (η) (Figure 5). We have remarked that a buffer electron mobility value under 100 $cm^2/V{\cdot}s$ would greatly affect the fill factor and efficiency. Defect densities are very important from the practical point of view in order to obtain required performance, especially for absorber and defect layers. We could compare different buffer and IDL materials from the point of view of their affinity. The optimal affinity interval (3.4–3.6 eV) for the buffer layer would determine the lower conduction band offset with the copper oxide absorber layer. The optimal affinity interval for IDL is larger (from 2.95 to 3.4 eV).

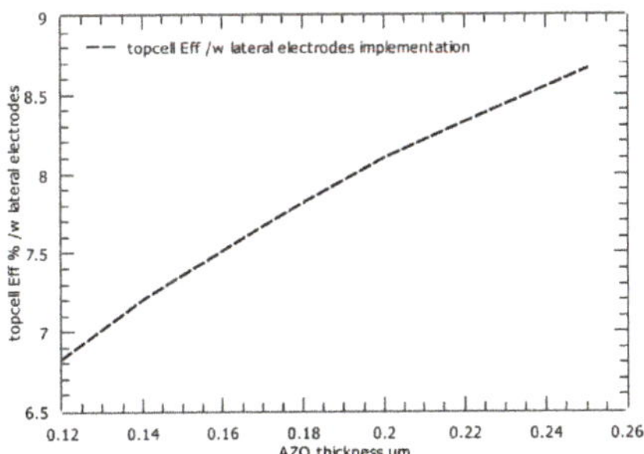

Figure 2. AZO (Al:ZnO) thickness vs. top cell efficiency.

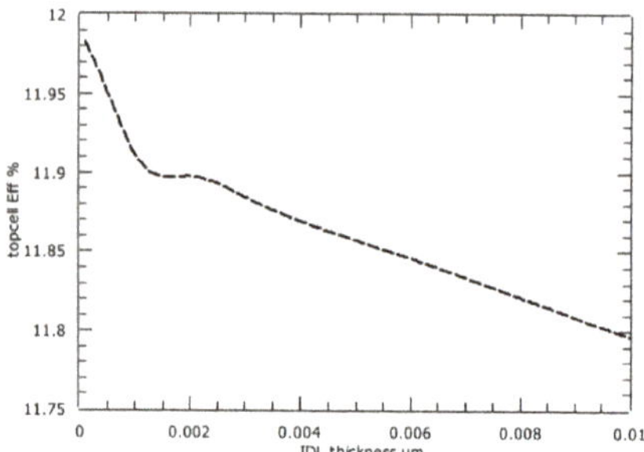

Figure 3. Interface defect layer (IDL) thickness vs. top cell efficiency.

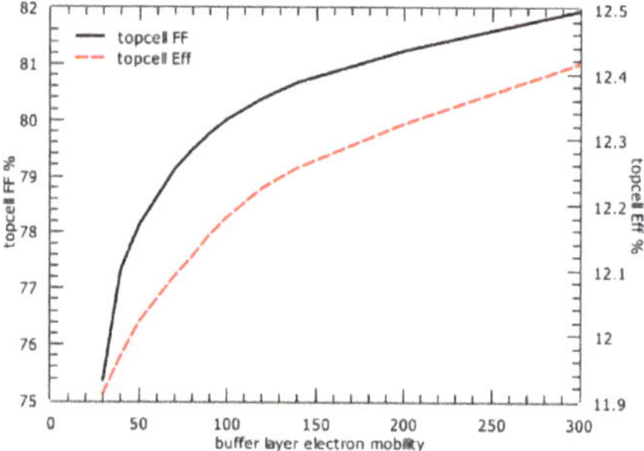

Figure 4. Buffer electron mobility ($cm^2/V{\cdot}s$) vs. top subcell fill factor and efficiency.

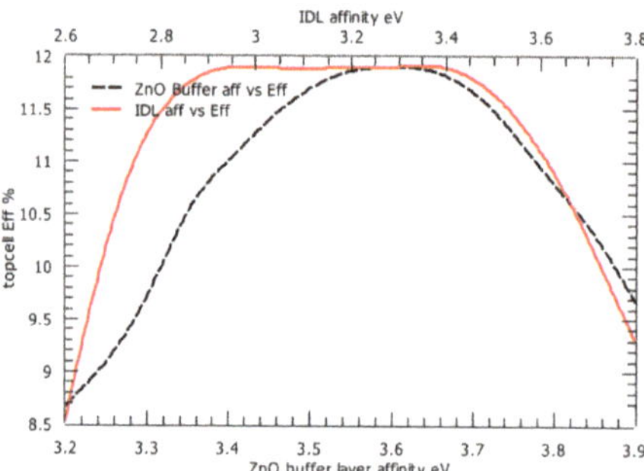

Figure 5. Buffer and IDL affinity vs. subcell efficiency.

A numerical model was adopted using the Quokka 2 software for the bottom subcell. The current-voltage (J-V) characteristics compared the experimental curve with the numerical modeled one (Figure 6).

The numerical model was used to fit to the experimental curve for control and prediction (Table 1) and to establish the essential electrical parameters for the silicone solar subcell.

A comparison between the experimental graph and Quokka modeled one for EQE (external quantum efficiency) is presented in Figure 7. Further improvement of the model would be required in the interval 750–1000 nm.

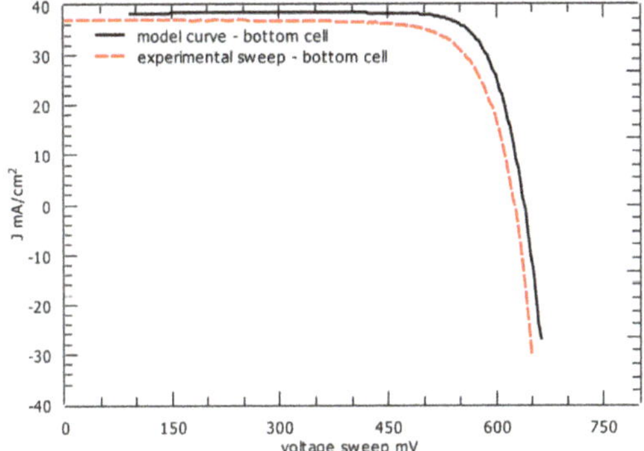

Figure 6. Experimental vs. Quokka 2 modeled J-V curve.

Table 1. Experimental J-V fit parameters for the c-Si subcell.

Parameter Name	Parameter Value
Open circuit voltage, V_{OC} (mV)	625.16
Fill Factor, FF (%)	76.83
Series resistance, R_S (ohm / cm^2):	0.196
Shunt resistance, R_{SH} (ohm / cm^2):	2382.8

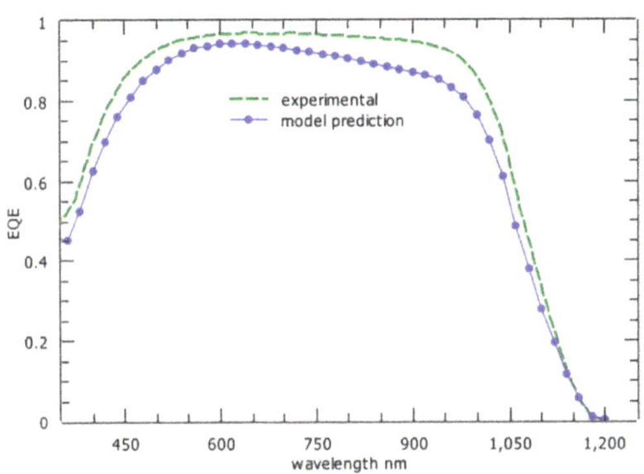

Figure 7. Experimental vs. Quokka modeled external quantum efficiency (EQE) curve.

The rather large current mismatch between the two subcells in tandem configuration, caused by the large band gap difference for the two absorbers, allows for individual optimization of the top and bottom subcells and favors the arrangement of a four-terminal configuration.

That is why it is possible to increase the conversion efficiency of the tandem heterojunction ZnO/Cu_2O silicone solar cell beyond the limit of conventional c-Si single-junction solar cell.

2.3. Optical Modeling and Simulation Results for the Top Subcell

In order to evaluate the optical characteristics of the device, we have used OPAL 2 software developed by PV (photovoltaic) Lighthouse [18]. OPAL 2 is a fast, reliable, and free online software tool that simulates the optics of advanced solar cells [19,20]. The optical modeling was performed for an AM1.5G solar spectrum and with experimentally determined complex refractive indices for the AZO and Cu_2O layers. The modeling was performed in the wavelength range from 300 nm to 800 nm.

Simulated optical functions were the normalized reflectance, absorptance, and transmittance as a function of wavelength for the studied device model, for the samples 1–3 of Cu_2O: N, to be discussed in the experimental section. The simulations are presented in Figure 8. For these simulations, a 100 nm thick AZO layer, a 2 µm thick Cu_2O layer, and a 50 nm thick Cu_2O:N layer were implemented in the model. A standard thickness of 50 nm was set for samples 1–3 of Cu_2O:N used in the solar cell structure.

The maximum reflectance in the 575–800 nm wavelength range for Sample 2 is a bit lower than the maximum values for Samples 1 and 3, as can be seen in Figure 8a. This result could prove useful for reflectance loss optimization.

The absorptance is high for all samples in the 300–525 nm wavelength range (with an average value of 90%). It decreases abruptly in the 525–600 nm range and tends to low values (<10%) in the 600–800 nm range, as seen in Figure 8b.

Transmittance was plotted concurrently for all three samples, as shown in Figure 8c. The transmittance suggests that Samples 1 and 3 yield a better transmission starting from the 560 nm wavelength. The cutoff wavelengths from the transmittance plot are 595 nm, 620 nm, 650 nm, 690 nm, 730 nm, 775 nm, and so on, a periodic spectral pattern being observed. The same approach can be taken for the reflectance plot. For the absorptance plot, the only significant cutoff wavelength is 590 nm. Based on the transmittance data, Tauc plots [20] for Samples 1–3 were calculated and presented in Figure 9.

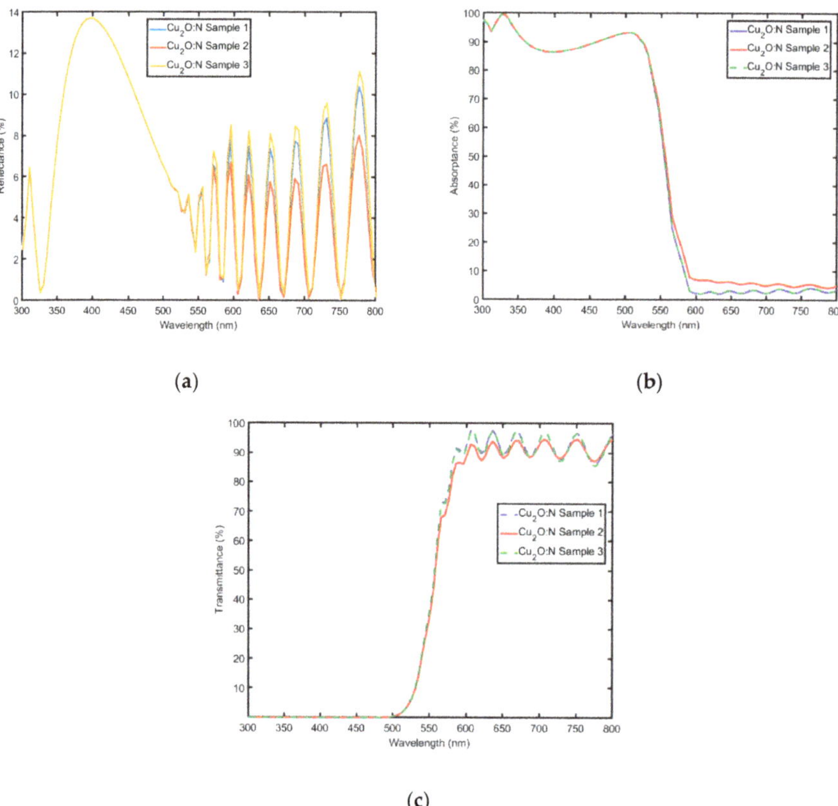

Figure 8. Percentage (**a**) reflectance, (**b**) absorptance, and (**c**) transmittance as a function of wavelength for the metal oxide structure.

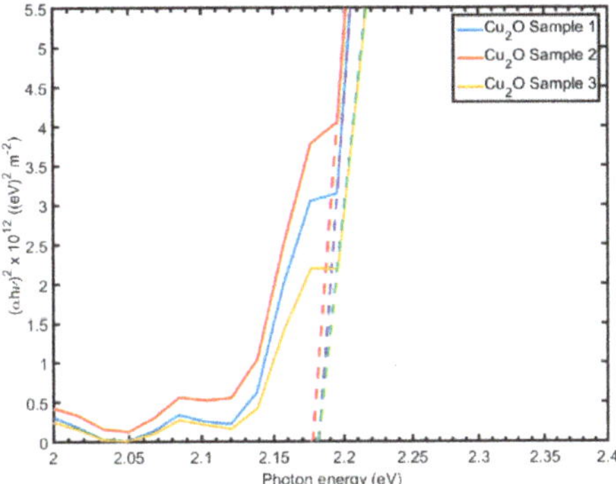

Figure 9. Tauc plots of Samples 1, 2, and 3 deposited on quartz substrate.

Tauc plots indicate an optical bandgap around 2.18 eV for Sample 1, 2.17 eV for Sample 2, and around 2.18 eV for Sample 3, which is in agreement with the ellipsometry measurements (see experimental section).

2.4. Relevance of the Best Simulation Tools for Numerical Modeling and Optimization of Metal Oxide Solar Devices

The simulation methodology could be used for design and material performance optimization. Two simulators were considered to be the best for the two subcells; Silvaco is the optimum tool for the top metal oxide solar subcell and absorbs the high energy part of sunlight, while PC1D is the most efficient tool for the bottom silicone solar subcell and absorbs the low energy part of sunlight. The simulated device with optimized parameters could be validated by comparing the numerical modeled results and performance with the experimental ones offered by an industrial manufactured solar cell with identical parameters.

3. Experimental Investigation of N-Doped Cu_2O Absorber Layer

3.1. Cu_2O Thin Film Fabrication and Overall Characterization

Cu_2O thin films were deposited on quartz substrates by direct current (DC) magnetron sputtering system (Semicore Triaxis) on 1 cm × 1 cm quartz substrates at 400 °C by varying the N_2 and Ar/O_2 flow [21]. Samples 1 and 2 were deposited at 400 °C with nitrogen flows 1 and 3 sccm (standard cubic centimeters per minute), respectively, while the Ar/O_2 flow was maintained constant. Sample 3 was deposited at 400 °C with Ar/N_2 flow of 1 sccm while the O_2 flow was fixed. The pre-sputtering time was 15 min, to avoid any contamination. The quartz substrates were cleaned in piranha solution and rinsed in deionized water, blown dry with nitrogen, and loaded into the deposition chamber. The same deposition time was used for all samples. The deposition power was fixed at 100 W. The base pressure before deposition was below 5×10^{-7} Torr while the sample stage was rotated at a constant speed of 12 rpm during deposition. The Hall effect measurements (LakeShore 7604) using the van der Pauw method were carried out to determine the hole mobility, resistivity, and hole carrier density.

The hole carrier density of as-deposited Sample 0 is ~10^{15} cm^{-3} with low mobility (~13 cm^2/Vs) and resistivity (~200 Ωcm), respectively. The hole carrier density for Samples 1 and 2 varied from ~10^{17}–10^{19} cm^{-3} with low carrier mobility (~0.1–1 cm^2/Vs) and resistivity (3–13 Ωcm), respectively, while Sample 3 has carrier density ~10^{16} cm^{-3} with resistivity ~54 Ωcm and carrier mobility ~1.5 cm^2/Vs. A reduced resistivity was achieved in the N-doped samples (1, 2, and 3) compared to the undoped Sample 0. This may reflect the presence of N in the DC Cu_2O film magnetron sputtering process influencing the electrical characteristics of solar cell as mentioned previously.

The process characteristics and layer structure of Cu_2O:N thin film samples, deposited by DC magnetron sputtering on quartz substrate, are presented in Table 2.

Table 2. Process characteristics and layer structure of Cu_2O:N thin films.

Sample Name	Gas Flow Conditions	Grains Size (nm)
Sample 0	Ar/O_2 was fixed at 42.5/7.5 sccm	20–90
Sample 1	N-doped 1 sccm (Ar/O_2 was fixed at 42.5/7.5 sccm)	20–48
Sample 2	N-doped 3 sccm (Ar/O_2 was fixed at 42.5/7.5 sccm)	30–54
Sample 3	N-doped 1 sccm (Ar was varied at 41.5 sccm, O_2 was fixed at 7.5 sccm)	40–88

3.2. *Morphological and Structural Characterization of the Cu_2O:N Films*

The surface morphology of the Cu_2O:N thin films was studied by SEM with the Quanta Inspect F 50 scanning electron microscope. Topographic AFM imaging was also performed by AFM with the Veeco Innova atomic force microscope, in the tapping mode, with 2.5 μm/s scanning speed. AFM images have a resolution of 512 × 512 pixels. The software used for image analysis was SPM Lab Analysis v.7.0 (Version 7.0, Veeco Innova, Bruker-Milton, ON, Canada, Santa Barbara-CA, USA.)

3.2.1. Scanning Electron Microscopy (SEM)

Experimental measurements for the determination of surface morphology were performed on Cu_2O:N thin films deposited by the DC magnetron sputtering on the quartz substrate under different gas flow conditions. Figure 10 shows the SEM images of four samples (see Table 2): one undoped Cu_2O film (Sample 0) and three Cu_2O:N thin film samples (Samples 1–3).

The grain size in the deposited layers is listed in Table 2 and compared with undoped Cu_2O film (Sample 0). With the increase of the N_2 flow, the films display a uniform layer surface with larger particle size (Samples 1 and 2). A varied Ar pressure induces a non-uniform particle size distribution (Samples 1 and 3). The particle size distribution of the undoped Cu_2O layer is non-uniformly distributed in the range 20–90 nm. As result, the presence of N_2 and an Ar/O_2 fixed flow improves the size uniformity of the deposited Cu_2O layers.

Figure 10. *Cont.*

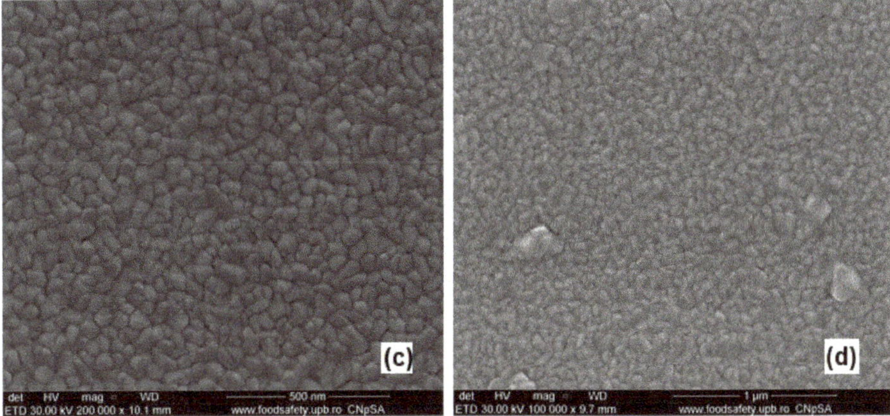

Figure 10. SEM images of Cu_2O thin films deposited on quartz substrate: (**a**) Sample 0, (**b**) Sample 1, (**c**) Sample 2, and (**d**) Sample 3.

3.2.2. Atomic Force Microscopy (AFM)

The grain size, including surface roughness of Cu_2O:N thin film deposited on quartz substrate, was measured by topographic AFM imaging. The surface morphologies for each of the N-doped thin film samples compared with Sample 0 are shown in Figure 11 as 2D AFM images, along with their corresponding surface profile histograms, and in Figure 12 as 3D AFM images. The root-mean-square surface roughness (R_{RMS}) for each sample was extracted from their AFM images [22,23].

The histograms represent the statistical distribution of the height surface profile of the Cu_2O samples derived from the AFM images and show a higher distribution in the first half of the height range for Samples 1 and 2. However, the height profile distribution of Sample 3 shows less roughness. This is related to the deposition conditions and in accordance with the R_{RMS} parameter calculations in Table 3, which shows a lower R_{RMS} for this sample. According to the statistical distribution of the surface height profile, Samples 1, 2, and 3 have the highest frequency at a surface height of 26 nm, 25 nm, and 17 nm, respectively. For validation, R_{RMS} was derived from the histograms based on the 2D AFM images, processed in MATLAB, and presented in Table 3.

Table 3. R_{RMS} of Cu_2O thin films and height distribution standard deviation, derived from histograms of 2D AFM images.

Sample Deposited on Quartz	R_{RMS} (nm)	Height Distribution Standard Deviation (nm)
Sample 0	6.3	7.8
Sample 1	4.7	5.8
Sample 2	4.2	4.5
Sample 3	3.9	4.2

According to Table 3, the roughness deviation values follow the same trend as the R_{RMS} values, where Sample 1 has the highest deviation and Sample 3 presents the lowest one. Additionally, the undoped Cu_2O film (Sample 0) presents a greater roughness than the doped Samples 1–3.

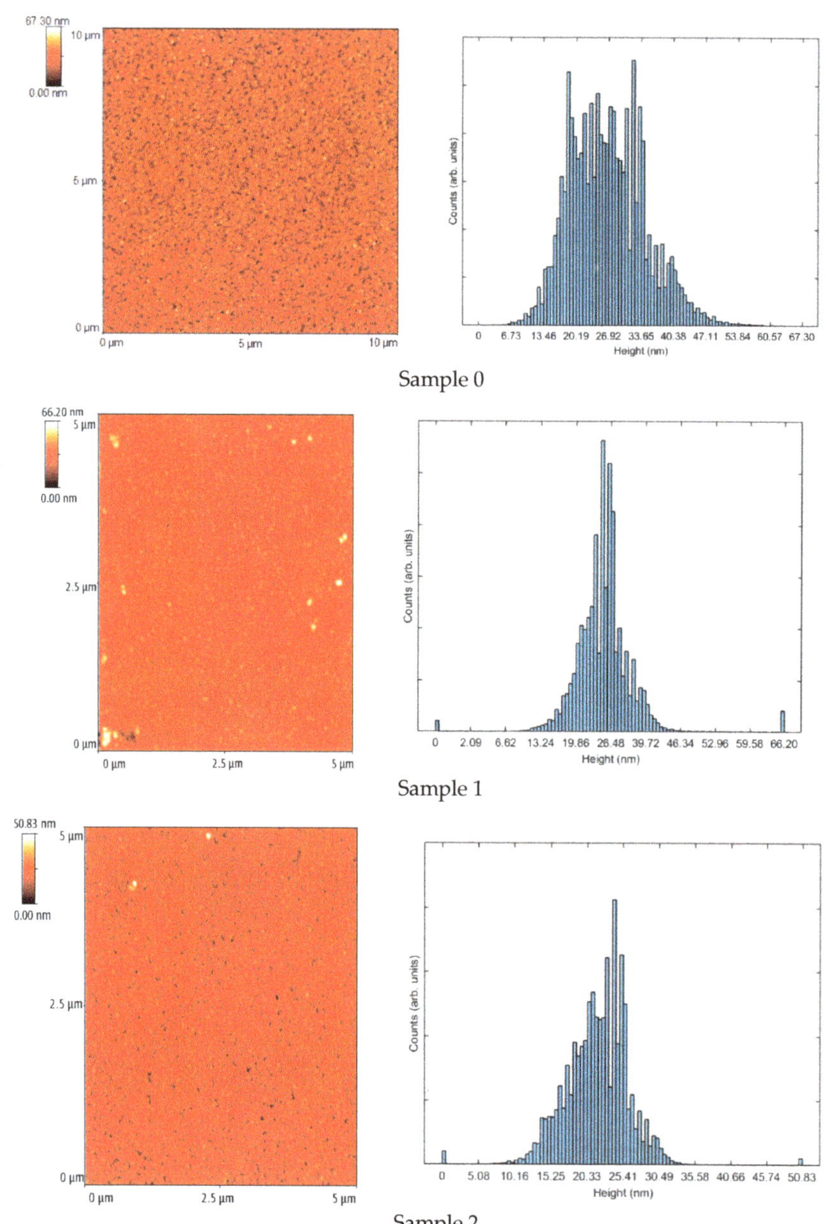

Figure 11. *Cont.*

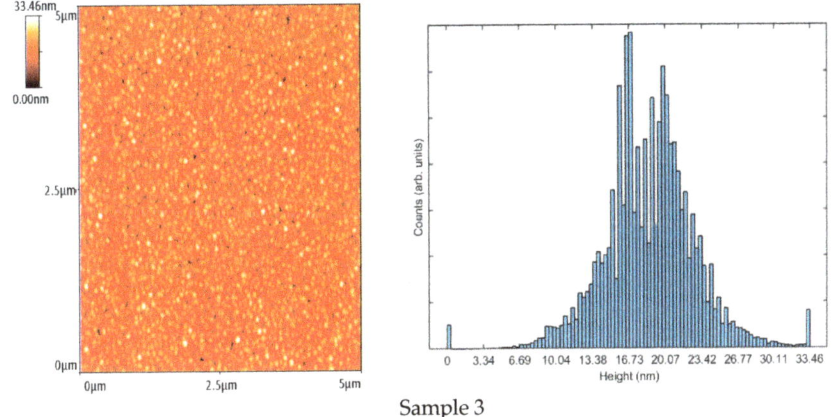

Sample 3

Figure 11. 2D AFM images of Samples 0–3 and their corresponding histograms.

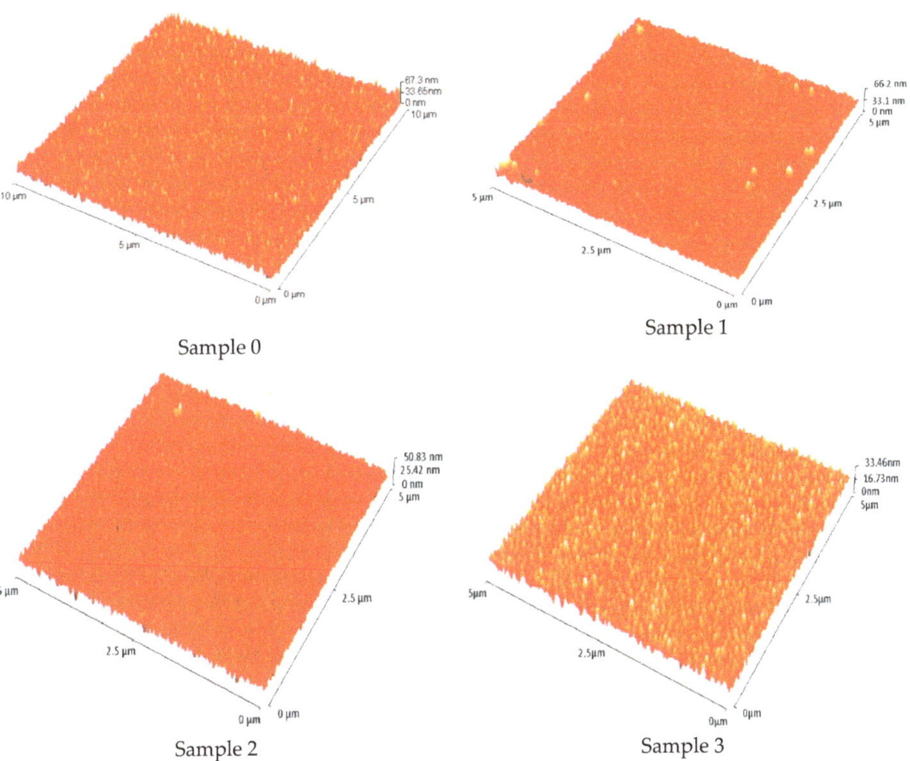

Figure 12. 3D AFM images of Samples 0–3.

3.2.3. Structural Characterization by XRD

The XRD pattern was obtained by a Bruker AXS D8 Discover Diffractometer, using Cu Kα-radiation and a Bragg–Brentano configuration. The background and Kα2 were stripped from the XRDs using the EVA software. The dominant XRD peaks were recorded in 2-theta range from 30–60°.

Figure 13 displays the XRD patterns of all samples, recorded in the range 30–60°. XRD patterns show that all films were dominated by a (200) reflection peak at 42.61° and a (111) reflection peak corresponding to 36.45°. The XRD pattern is not significantly influenced by nitrogen doping [24]. The Cu_2O peaks were determined by comparing experimental XRD peak patterns with the standard Powder Diffraction cards (ICDD patterns: 01-071-3645) [25].

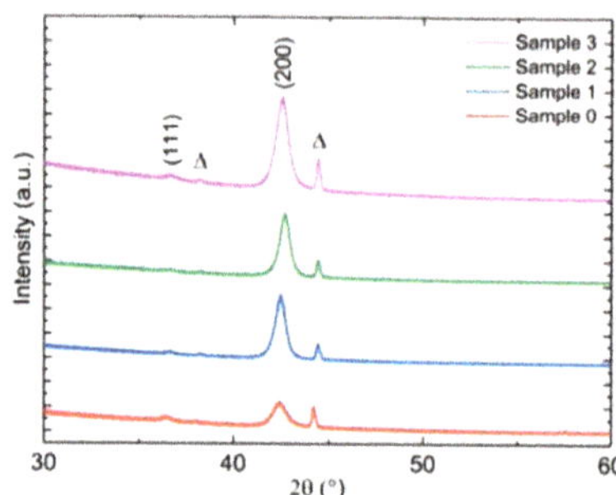

Figure 13. XRD patterns of Samples 0–3 in the range 30–60°, where Δ represents the background peak.

XRD patterns of the Cu_2O:N films prepared under various N_2 flow rates, together with the data of the Cu_2O reference films, show highly oriented (111) planes. For the Cu_2O:N samples, the peak intensity of the (111) plane decreases significantly with increasing of N_2 flow rate, and (200) planes show up in the heavy N-doping range [26].

3.3. *Optical Characterization*

Optical characterization involved two tools: FTIR spectroscopy and spectroscopic ellipsometry. FTIR spectroscopy analysis was carried out using a Perkin Elmer type Spectrum 100 spectrometer, with UATR accessory—PIKE GladiATR, in the wavenumber range 430–2000 cm^{-1} with a spectral resolution of 0.4 cm^{-1}. The refractive index and extinction coefficient of Cu_2O:N thin film were determined by spectroscopic ellipsometry, in the wavelength range from 190 to 2100 nm using an UVISEL spectroscopic ellipsometer from HORIBA Jobin Yvon. A model fit to the measured ellipsometry parameters was obtained using DeltaPsi 2.6 software (Version 2.6, HORIBA, Jobin Yvon, Palaiseau, France).

3.3.1. FTIR Spectroscopy

The FTIR transmission spectra obtained for Cu_2O:N thin films are shown in Figure 14 and the transmission peaks are listed in Table 4.

Table 4. Transmission peaks for Cu_2O:N thin films.

Sample Name	νCu_2O (cm^{-1})
Sample 1	608
Sample 2	616
Sample 3	615

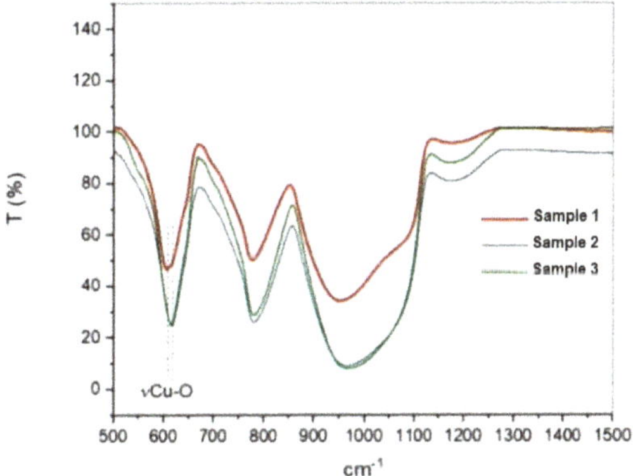

Figure 14. FTIR transmission spectra for Cu_2O:N thin film: Sample 1, Sample 2, and Sample 3.

The FTIR spectra of all samples deposited on quartz displayed the peaks corresponding to the vibrational modes of Si-O belonging to the quartz substrate (780 cm^{-1}, broadband centered at 980 cm^{-1} with a shoulder at 1080 cm^{-1}) [26,27]. The band located at 608–616 cm^{-1} corresponds to the stretching vibrational mode of Cu-O in Cu_2O phase. There is an increasing trend in the intensity of the Cu-O absorption band from Sample 1 to Sample 3, suggesting an increased concentration of N_2. This result is in agreement with the recent theoretical predictions that nitrogen mainly substitutes Cu in the molecular form, (N_2) Cu, in N-doped Cu_2O rather than as substitutional N on the O site, (NO) [5]. The decrease in resistivity, coherently with the increase in N doping concentration, was explained by annealing-driven substitution of oxygen atoms with implanted N atoms. The N doping may induce any definite variation in energy band structure of Cu_2O films [25].

In general, the photoluminescence (PL) spectra of Cu_2O films contain two main signals ranging from 450 to 650 nm, which are considered to be derived from free excitons and the bound excitonic region [28].

3.3.2. Spectroscopic Ellipsometry (SE)

The ellipsometry parameters of Cu_2O:N thin films deposited on quartz substrate were investigated with a resolution of 27 points in the range 1.5–4.1 eV (with step 0.1 eV). The investigated parameters were the band gap and the surface roughness [29,30] and were listed in Table 5.

Table 5. Investigated (SE) parameters of Cu_2O:N thin films.

Sample Name	Sample 1	Sample 2	Sample 3
Band gap (eV):	2.17	2.14	2.17
Surface roughness (nm):	15.9	19.6	17.2

By the Tauc–Lorentz oscillator methodology [31], modeling was performed to determine the band gap, surface roughness, refractive index (n), and the extinction coefficient (k) in the range 300–800 nm (from visible to near infrared) for Cu_2O thin films deposited on quartz substrates.

The model described a layer of Cu_2O deposited on the quartz substrate with a hard surface layer. The surface roughness was modeled by mixing the optical constants of the Cu_2O material surface and the air voids (on average, 90% Cu_2O and 10% air voids).

It is noticed that the surface roughness of Cu_2O film increases with the level of N_2 pressure (Sample 1 and Sample 3 vs. Sample 2), from 15.9 nm to 19.6 nm (Table 5).

The presence of N_2 during deposition of Cu_2O films on quartz substrate produces small changes over n and k in the range 300–600 nm for Samples 1–3 and 300–400 nm for Sample 2, as can be seen in Figure 15.

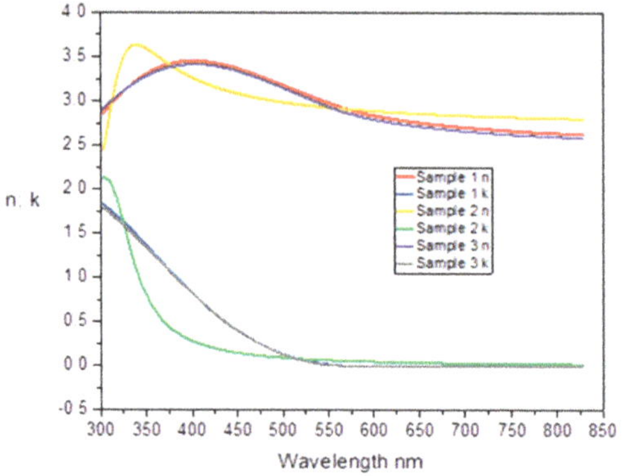

Figure 15. Graphs of n and k coefficients for Samples 1, 2, and 3 of Cu_2O:N thin films; λ is wavelength.

The changes of n(λ) (yellow curve) and k(λ) (green curve) coefficients observed in Figure 15 for Sample 2 (the highest N-doped sample) are associated with a structural change in Cu_2O induced by nitrogen doping. This can be a promising method for modifying the optical properties of the metal oxide [28,32] The crystal size decreases from approximately 23 nm to approximately 15 nm with increasing the N-doping concentration in the small N-doping range, where as it is almost unchanged (18–20 nm) in the heavy N-doping range [26].

For example, undoped Cu_2O films were deposited on quartz substrates by the RF magnetron sputtering system with a Cu_2O target (99.99%, purity); with the Ar flow rate at 20 sccm, the average transmittances in the visible range were 40–60%, and the band gap of Cu_2O films was approx. 18 eV [29].

3.4. General Comments on Experimental Investigation and Prospects of Heterojunction Cuprous Oxide Solar Cells with Nitrogen

The experimental investigation of N-doped Cu_2O thin films show they have good potential as an absorber layer for photovoltaic applications:

- The doped samples are characterized by higher whole carrier densities, lower carrier mobilities, and lower resistivities in comparison with the undoped sample;
- The SEM images showed the improvement in the size uniformity of the deposited Cu_2O layers due to N presence;
- The AFM measurements determined that the roughness is slightly influenced by the N doping, which could affect the electrical properties of solar cells. The histograms derived from the AFM images showed the surface profile height for the studied Cu_2O:N samples and emphasized their roughness deviation;
- The X-ray analysis of Cu_2O films put in evidence that all such films are polycrystalline and nitrogen doping does not show any significant change in the XRD pattern. A reduced resistivity was achieved in the N-doped samples compared to the undoped sample;
- The FTIR spectra of the samples displayed peaks corresponding to the vibrational modes of Si-O belonging to the quartz substrate;
- The SE analysis stressed the N influence over the optical properties of the Cu_2O thin films. The changes of refraction and extinction coefficients were given by the Cu_2O:N thin films associated with the structural change in Cu_2O induced by N-doping.

At the same time, it could be remarked that the N-doping can lead to the following on the heterojunction copper oxide solar cells with nitrogen:

- regulation of the band gap enabling increased transmittance;
- improved carrier lifetime and transport;
- increased conversion efficiency of the solar cell device.

All these results highlight the N-doped cuprous oxide heterojunction solar cell as a promising candidate for high efficiency solar devices.

4. Reliability Study

4.1. Methodological Tools

Definition of solar cell reliability: probability of device performance to fit to specific performance parameters for clear operational conditions without deviation outside determined tolerance limits.

There are different accelerated tests to be done regarding the reliability of a solar cell:

(1) Qualitative accelerated test: offers information regarding the solar cell malfunctions or failure ways;
(2) High accelerated life test (HALT) offers information regarding the solar cell failure mechanisms and average lifetime;
(3) High accelerated stress screening (HASS) offers information regarding the solar cell operating time and failure rate within the operating temperature range

In order to perform an accurate reliability analysis of the solar cell, the SYNTHESIS simulation tools developed by ReliaSoft company [33] were used, namely: (1) *the Weibull++ tools*, which performs life data analysis using multiple lifetime repartitions with a reliability-oriented interface; (2) *the ALTA tool*, which allows a quantitative analysis of accelerated life tests.

4.2. Solar Cells Thermal Stability Characteristics

Three main parameters can be studied using the Weibull statistical repartition: (1) *the solar cell lifetime* in different conditions: (2) *the degradation degree* based on the temperature influence; and (3) *number of defects (failures).*

In order to make the most accurate predictions for the lifetime of the analyzed solar cells, the Weibull statistical repartition method (Figure 16a) of the Weibull++ tool was employed [33].

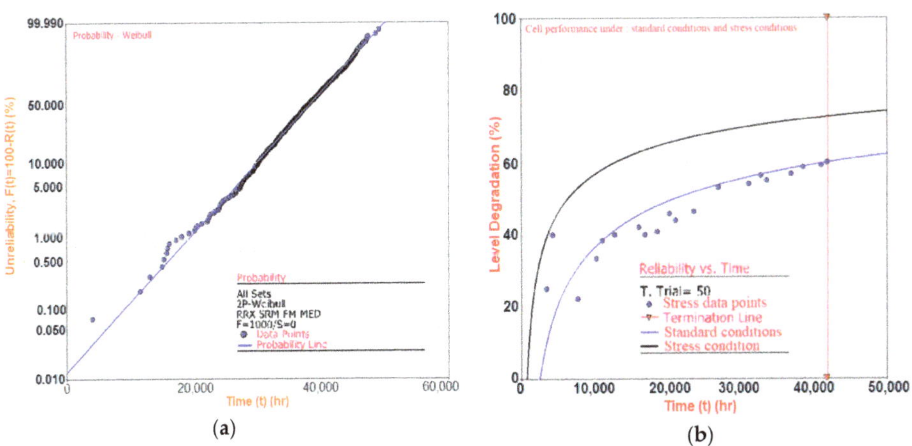

Figure 16. (**a**) Weibull distribution for the studied solar cell lifetime; (**b**) solar cell performance and degradation degree of the analyzed solar cell under standard and stress conditions.

The degradation degree of the analyzed solar cell can be characterized by the temperature influence on the device [34–36] and is obtained for two cases: (1) under stress conditions (black curve) and (2) under standard conditions (blue curve) (Figure 16b).

Based on the stress results of the HASS test presented in Figure 16b, it is possible to determine, by numerical simulation using *the ALTA tool*, the limiting value of specific solar cell operating time in normal conditions, as seen in Figure 17a. The HASS test also establishes the probable rate of solar cell failure within the operation temperature range of 50–100 °C (Figure 17b). It is remarked that the expected failures rate dependence increases strongly with temperature.

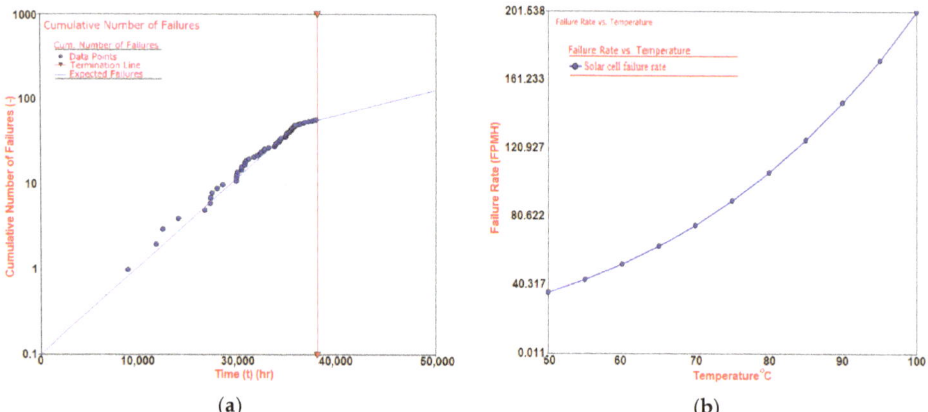

Figure 17. (**a**) Cumulative number of failures as function of time (using ALTA module). (**b**) Baseline survival of solar cell simulation (using ALTA module) in temperature and stress conditions.

The simulated results obtained from this reliability analysis consist both in identifying the solar cell failure modes, as well as to predict the solar cell lifetime in normal operation conditions. At the same time, these simulations enable evaluation of the degradation time and number of defects for the analyzed solar cell. The number of defects increases alarmingly starting at 20,000 operating hours, and by the end of the accelerated HASS test, the solar device would fail (Figure 17a).

4.3. Further Work

The simulation tools used in this work, namely Weibull++ and ALTA, have limited capabilities for analyzing different types of advanced solar cells based on different materials. That is why new simulation tools must be considered for a further work dedicated to reliability studies and experimental analyses of metal oxide solar cells. The thermal stability of such devices is essential during solar cell operation. It could be noted that the Cu_2O:N layer is very stable because its annealing for 90 min does not degrade its nitrogen-doping, as reported by Ye F. et al. [37].

5. Conclusions

The optimization of a silicon tandem heterojunction solar cell with a Cu_2O absorber layer has been analyzed by numerical modeling based on its splitting into two subcells: a ZnO/Cu_2O top subcell and a c-Si bottom one. The buffer layer thickness and affinity are essential parameters for implementation of such devices defined by a low conduction band offset.

The optical characteristics of a Cu_2O heterojunction solar cell were obtained by numerical modeling, allowing the evaluation of optical losses and determination of maximum absorptance of the device.

A comprehensive characterization (morphological, structural, and optical) of the copper oxide layer with nitrogen doping enables us to propose the metal oxide heterojunction solar cell with nitrogen as a promising candidate for advanced high efficiency solar cells.

The main reliability parameters of metal oxide solar cells, namely the failure rate and degradation degree, were calculated and further work will stress the influence on electrical and optical properties of these devices from the thermal stability point of view.

Author Contributions: Conceptualization, L.F., I.C., and Ø.N.; methodology, L.F., I.C., R.K., and D.C.; formal analysis, L.F.; investigation, L.F., I.C., R.K., C.V., L.B., and D.S.; funding acquisition, L.F. and I.C.; software, S.F. and D.C.; validation, L.F., I.C., Ø.N., J.P.C., and D.C.; resources, I.C., R.K., and L.F.; data curation, S.F.; writing—original draft preparation, L.F., I.C., and S.F.; writing—review and editing, L.F., and I.C.; visualization, L.F. and I.C.; supervision, L.F., I.C., and J.P.C.; project administration, J.P.C., Ø.N., E.M., L.F., and I.C. All authors have read and agreed to the published version of the manuscript.

Funding: This research received no external funding.

Acknowledgments: This work was supported by the project "Multiscale in modelling and validation for solar photovoltaics" (Multiscale Solar) MP1406, 2015–2019, for IPVF (Institute Photovoltaïque de l'Île de France)-GeePs, PUB (Polytechnic University of Bucharest) and INOE, funded by the European Commission through COST program. This research activity was conducted under the research project "High-performance tandem heterojunction solar cells for specific applications (SOLHET)", for IFE, UiO, PB and INOE funded by the Research Council of Norway (RCN), project no. 251789, and the Romanian Executive Agency for Higher Education, Research, Development and Innovation Funding (UEFISCDI), contracts no. 34/2016 and no. 35/2016, through the M-ERA.NET program, and project no. PN 19_18.01.01/2019, Ad. Act. 2020, for INOE, funded by Ministry of Research and Innovation (MCI), through the NUCLEU program and institutional project 19PFE/2018, Stage IV/2020.

Conflicts of Interest: The authors declare no conflict of interest.

References

1. Wang, Y.; Miska, P.; Pilloud, D.; Horwat, D.; Mücklich, F.; Pierson, J.F. Transmittance enhancement and optical band gap widening of Cu_2O thin films after air annealing. *J. Appl. Phys.* **2014**, *115*, 073505. [CrossRef]
2. Jung, Y.S.; Choi, H.W.; Kim, K.H. Properties of p-type N-doped Cu_2O thin films prepared by reactive sputtering. *Jpn. J. Appl. Phys.* **2014**, *53*, 11RA10. [CrossRef]
3. Siddiqui, H.; Parra, M.R.; Pandey, P.; Singh, N.; Qureshi, M.S.; Haque, F.Z. A review: Synthesis, characterization and cell performance of Cu_2O based material for solar cells. *Orient. J. Chem.* **2012**, *28*, 1533–1545. [CrossRef]
4. Wang, Y.; Ghanbaja, J.; Horwat, D.; Yu, L.; Pierson, J.F. Nitrogen chemical state in N-doped Cu_2O thin films. *Appl. Phys. Lett.* **2017**, *110*, 131902. [CrossRef]
5. Sberna, P.M.; Crupi, I.; Moscatelli, F.; Privitera, V.; Simone, F.; Miritello, M. Sputtered cuprous oxide thin films and nitrogen doping by ion implantation. *Thin Solid Film.* **2016**, *600*, 71–75. [CrossRef]
6. Mitroi, R.M.; Ninulescu, V.; Fara, L. Tandem solar cells based on Cu_2O and c-Si subcells in parallel configuration: Numerical simulation. *Int. J. Photoenergy* **2017**, *2017*, 1–6. [CrossRef]
7. Takiguchi, Y.; Miyajima, S. Device simulation of cuprous oxide heterojunction solar cells. *Jpn. J. Appl. Phys.* **2015**, *54*, 112303. [CrossRef]
8. Nordseth, O.; Fara, L.; Kumar, R.; Foss, S.E.; Dumitru, C.; Muscurel, V.F.; Drăgan, F.; Crăciunescu, D.; Bergum, K.; Haug, H.; et al. Electro-optical modeling of a ZnO/Cu_2O subcell in a silicon-based tandem heterojunction solar cell. In Proceedings of the 33rd European Photovoltaic Solar Energy Conference and Exhibition, Amsterdam, The Netherlands, 25–29 September 2017; pp. 172–177. [CrossRef]
9. Nordseth, O.; Haug, H.; Kumar, R.; Bergum, K.; Drăgan, F.; Craciunescu, D.; Fara, L.; Chilibon, I.; Monakhov, E.; Foss, S.E.; et al. Performance optimization of a four-terminal Cu_2O/c-Si Tandem heterojunction solar cell. In Proceedings of the 35th European Photovoltaic Solar Energy Conference and Exhibition, Bruxelles, Belgium, 24–28 September 2018; pp. 29–34. [CrossRef]
10. Dumitru, C.; Fara, L.; Nordseth, O.; Chilibon, I.; Kumar, R.; Svensson, B.G.; Dragan, F.; Muscurel, V.; Craciunescu, D.; Sterian, P. Electro-optical analysis and numerical modeling of Cu_2O as the absorber layer in advanced solar cells. In Proceedings of the 2018 International Conference on Photovoltaic Science and Technologies (PVCon), Ankara, Turkey, 4–6 July 2018; pp. 1–7. [CrossRef]

11. Chilibon, I.; Fara, L.; Nordseth, O.; Kumar, R.; Svensson, B.G.; Sterian, P.; Dumitru, C.; Dragan, F.; Muscurel, V.; Vasiliu, C. Structural and electrical analysis of Cu_2O layers for solar cell application. *Ann. Acad. Rom. Sci.* **2018**, *11*, 53–60.
12. Chilibon, I.; Fara, L.; Nordseth, O.; Kumar, R.; Svensson, B.G.; Sterian, P.; Dumitru, C.; Dragan, F.; Muscurel, V.; Vasiliu, C. Characterization of Cu_2O thin films used in solar cells by fluorescence and FTIR spectroscopy. *Ann. Acad. Rom. Sci. Ser. Sci.* **2018**, *11*, 61–68.
13. Lloyd, M.A.; Siah, S.C.; Brandt, R.E.; Serdy, J.; Johnson, S.W.; Lee, Y.S.; Buinassisi, T. Intrinsic defect engineering of cuprous oxide to enhance electrical transport properties for photovoltaic applications. In Proceedings of the IEEE 40th Photovoltaic Specialists Conference (PVSC), Denver, CO, USA, 8–13 June 2014; Volume 2016, pp. 3443–3445.
14. Tolstova, Y.; Omelchenko, S.T.; Blackwell, R.E.; Shing, A.M. Polycrystalline Cu_2O photovoltaic devices incorporating Zn (O, S) window layers. *Sol. Energy Mat. Sol. Cells* **2017**, *160*, 340–345. [CrossRef]
15. Minami, T.; Nishi, Y.; Miyata, T. High-efficiency Cu_2O-based heterojunction solar cells fabricated using a Ga_2O_3 thin film as n-type layer. *Appl. Phys. Express* **2013**, *6*, 044101. [CrossRef]
16. Robertson, J.; Falabretti, B. Band offsets of high K gate oxides on III-V semiconductors. *J. Appl. Phys.* **2006**, *100*, 014111. [CrossRef]
17. Brandt, R.E.; Young, M.; Hejin, H.P.; Dameron, A.; Chua, D.; Lee, S.Y.; Teeter, G.; Gordon, R.; Buonassisi, T. Band offsets of n-type electron-selective contacts on cuprous oxide (Cu_2O) for photovoltaics. *Appl. Phys. Lett.* **2014**, *105*, 263901. [CrossRef]
18. PV Lighthouse. Available online: https://www.pvlighthouse.com.au/ (accessed on 20 January 2020).
19. Baker-Finch, S.C.; McIntosh, K.R. OPAL 2: Rapid optical simulation of silicon solar cells. In Proceedings of the 38th IEEE Photovoltaic Specialists Conference, Austin, TX, USA, 3–8 June 2012; pp. 265–271. [CrossRef]
20. Dumitru, C.; Muscurel, V.; Nordseth, O.; Fara, L.; Sterian, P. Optimization of electro-optical performance and material parameters for a tandem metal oxide solar cell. In Proceedings of the International Conference on Computational Science and Its Applications–ICCSA, Melbourne, Australia, 2–5 July 2018; pp. 573–582. [CrossRef]
21. Murali, D.S.; Kumar, S.; Choudhary, R.J.; Wadikar, A.D.; Jain, M.K.; Subrahmanyam, A. Synthesis of Cu_2O from CuO thin films: Optical and electrical properties. *AIP Adv.* **2015**, *5*, 047143. [CrossRef]
22. Bennett, J.M.; Mattson, L. *Introduction to Surface Roughness and Scattering*; Optical Society of America: Washington, DC, USA, 1989.
23. Nordseth, O.; Kumar, R.; Bergum, K.; Chilibon, I.; Fara, L.; Foss, S.E.; Monakhov, E. Nitrogen-doped Cu_2O thin films for photovoltaic applications. *Materials* **2019**, *12*, 3038. [CrossRef] [PubMed]
24. Lee, S.W.; Lee, Y.S.; Heo, J.; Siah, S.C.; Chua, D.; Brandt, R.; Kim, S.B.; Mailoa, J.P.; Buonassisi, T.; Gordon, R.G. Improved Cu2O-based solar cells using atomic layer deposition to control the Cu oxidation state at the p-n junction. *Adv. Energy Mater.* **2014**, *4*, 1301916. [CrossRef]
25. *JCPDS 78-2076 for Cu_2O, JCPDS 03-0879 for Cu_4O_3, JCPDS 48-1548 for CuO, and JCPDS 85-1326 for Cu*; International Centre for Diffraction Data-JCPDS: Newtown Square, PA, USA, 1996.
26. Nakano, Y.; Saeki, S.; Morikawa, T. Optical bandgap widening of p-type Cu_2O films by nitrogen doping. *Appl. Phys. Lett.* **2009**, *94*, 022111. [CrossRef]
27. Reddy, M.H.P.; Pierson, J.F.; Uthanna, S. Structural, surface morphological, and optical properties of nanocrystalline Cu_2O and CuO films formed by RF magnetron sputtering: Oxygen partial pressure effect. *Phys. Status Solidi A* **2012**, *209*, 1279–1286. [CrossRef]
28. Ogwu, A.A.; Darma, T.H. A reactive magnetron sputtering route for attaining a controlled core-rim phase partitioning in Cu_2O/CuO thin films with resistive switching potential. *J. Appl. Phys.* **2013**, *113*, 183522. [CrossRef]
29. Su, J.; Niu, Q.; Sun, R.; An, X.; Zhang, Y. Investigation of Cu_2O films sputtered with ceramic target: Effect of RF power. *J. Optoelectron. Adv. M.* **2018**, *20*, 441–444.
30. Gong, J.B.; Dong, W.L.; Dai, R.C.; Wang, Z.P.; Zhang, Z.M.; Ding, Z.J. Thickness dependence of the optical constants of oxidized copper thin films based on ellipsometry and transmittance. *Chin. Phys. B* **2014**, *23*, 1–5. [CrossRef]
31. Gan, J.; Venkatachalapathy, V.; Svensson, B.G.; Monakhov, E.V. Influence of target power on properties of Cu_xO thin films prepared by reactive radio frequency magnetron sputtering. *Thin Solid Film.* **2015**, *594*, 250–255. [CrossRef]

32. Li, J.; Mei, Z.; Liu, L.; Liang, H.; Azarov, A.; Kuznetsov, A.; Liu, Y.; Ji, A.; Meng, Q.; Du, X. Probing defects in nitrogen-doped Cu_2O. *Sci. Rep.* **2014**, *4*, 7240. [CrossRef] [PubMed]
33. Reliability Engineering Software Products–ReliaSoft. Available online: https://www.reliasoft.com (accessed on 20 January 2019).
34. Sharma, V.; Chandel, S.S. Performance and degradation analysis for long term reliability of solar photovoltaic systems: A review. *Renew. Sustain. Energy Rev.* **2013**, *27*, 753–767. [CrossRef]
35. Guo, D.; Brinkman, D.; Shaik, A.R.; Ringhofer, C.; Vasileska, D. Metastability and reliability of CdTe solar cells. *J. Phys. D Appl. Phys.* **2018**, *51*, 15. [CrossRef]
36. Fara, L.; Craciunescu, D. Reliability analysis of photovoltaic systems for specific applications. In *Reliability and Ecological Aspects of Photovoltaic Modules*; Gok, A., Ed.; Intech Open: London, UK, 2020; pp. 79–92.
37. Ye, F.; Zeng, J.J.; Qiu, Y.B.; Cai, X.M.; Wang, B.; Wang, H.; Zhang, D.P.; Fan, P.; Roy, V.A.L. Deposition-rate controlled nitrogen-doping into cuprous oxide and its thermal stability. *Thin Solid Film.* **2019**, *674*, 44–51. [CrossRef]

Article

Analysis of Contact Reaction Phenomenon between Aluminum–Silver and p+ Diffused Layer for n-Type c-Si Solar Cell Applications

Cheolmin Park [1], Sungyoon Chung [2], Nagarajan Balaji [1], Shihyun Ahn [3], Sunhwa Lee [2], Jinjoo Park [4] and Junsin Yi [2,*]

1 Department of Energy Science, Sungkyunkwan University, Suwon 16419, Korea; cmpark8311@gmail.com (C.P.); balaji@skku.edu (N.B.)
2 College of Information and Communication Engineering, Sungkyunkwan University, Suwon 16419, Korea; sungyoun89@skku.edu (S.C.); ssunholic@skku.edu (S.L.)
3 Department of Electrical and Computer Engineering, University of Seoul, 163, Seoulsiripdae-ro, Dongdaemun-gu, Seoul 02504, Korea; shahn@uos.ac.kr
4 Major of Energy and Applied Chemistry, Division of Energy & Optical Technology Convergence, Cheongju University 298, Daeseong-ro, Cheongwon-gu, Cheongju-si, Chungcheongbuk-do, Cheongju 28503, Korea; jwjh3516@cju.ac.kr
* Correspondence: junsin@skku.edu; Tel.: +82-31-290-7139

Received: 20 July 2020; Accepted: 31 August 2020; Published: 2 September 2020

Abstract: In this study, the contact mechanism between Ag–Al and Si and the change in contact resistance (Rc) were analyzed by varying the firing profile. The front electrode of an n-type c-Si solar cell was formed through a screen-printing process using Ag–Al paste. Rc was measured by varying the belt speed and peak temperature of the fast-firing furnace. Rc value of 6.98 mΩ-cm^{-2} was obtained for an optimal fast-firing profile with 865 °C peak temperature and 110 inches per min belt speed. The contact phenomenon and the influence of impurities between the front-electrode–Si interface and firing conditions were analyzed through scanning electron microscopy (SEM) and energy-dispersive X-ray spectroscopy (EDS). The EDS analysis revealed that the peak firing temperature at 865 °C exhibited a low atomic weight percentage of Al (0.72 and 0.36%) because Al was involved in the formation of alloy of Si with the front electrode. Based on the optimal results, a solar cell with a conversion efficiency of 19.46% was obtained.

Keywords: metallization; contact formation; Ag/Al paste; p+ emitter; N-type bifacial solar cells

---

1. Introduction

The optimization of the characteristics of the front electrodes of solar cells is important in two major aspects—improving conversion efficiency and reducing production costs, which are the main aims while fabricating solar cells. In the solar cell industry, screen printing technologies occupy most of the market share of the metallization of front and rear sides [1–4]. Based on the advantages of manufacturing cost and throughput, the technology of the screen-printing process has continuously improved, and the technological improvement in conversion efficiency can be largely divided into two aspects. The first is minimizing the shading loss by applying fine-line printing to absorb as much light as possible [5], and the second is minimizing the series resistance to reduce electrical loss when the carriers generated by the absorbed light are collected at the electrode [6]. For fine-line printing, techniques such as changing the emulsion thickness and the application of a knotless screens were applicable to both p- and n-type substrates regardless of the substrate type [7,8].

However, when decreasing the series resistance (specifically contact resistance (Rc)), changes in substrate type are significant factors. For p-type substrates, an emitter is formed using phosphorus, and phosphorus atoms penetrate the phosphorus silicate glass (PSG) layer and do not out-diffuse. As a result, the phosphorus atoms accumulate on the surface; thus, the carrier concentration on the surface is the highest [9,10]. In contrast, for n-type substrates, boron atoms penetrate the borosilicate glass (BSG) and diffuses out, thereby forming a "deficient layer" with a relatively low boron concentration on the surface [11]. When a deficient layer is formed, there are limitations to reducing the Rc when using Ag paste. This is because the Rc (resistivity) and emitter surface concentration have the following formula [12]:

$$\rho_c = \rho_{co} exp\left(\frac{2\Phi_B}{\hbar}\sqrt{\frac{\varepsilon_s m^*}{N}}\right) ohm - cm^2 \tag{1}$$

where ρ_{co} is a constant dependent on the metal and semiconductor. The specific contact resistivity (ρ_c) primarily depends on the metal-semiconductor work function (Φ_B), doping density (N), and effective mass of the carrier (m^*). Therefore, the paste of the n-type substrate solar cell contains a certain amount of Al to compensate for the deficient layer. It is similar to the formation of the local back surface field (LBSF) of the p-type passivated emitter and rear cell (PERC) structure, where Al reacts with Si and heat to form a p+ region when an alloy is formed. A small amount of Al in the front paste aids in improving the properties of the front contact.

In this study, the contact formation between p+ diffused Si layers and Ag–Al paste with respect to the firing conditions were qualitatively and quantitatively analyzed using scanning electron microscopy (SEM) and energy-dispersive X-ray spectroscopy (EDS). Based on the analysis, we aimed at describing the bonding pattern according to the temperature using a schematic diagram. Finally, the output characteristics of the n-type c-Si solar cell fabricated under optimized conditions were analyzed.

2. Experimental Details

Figure 1 shows the schematic of the solar cell structure and process flow used in this study. Here, 200-μm thick n-type solar-grade Czochralski (Cz-Si) grown wafers with a crystal orientation of (100) and resistivity of 1.5 ohm-cm were used as the starting material for the sample to be analyzed and fabricated into a solar cell. To reduce the reflectance of the Si wafer, the sample was textured using 2% NaOH and 8.75% isopropanol at a temperature range of 84 to 86 °C. The textured wafers were doped with boron using a thermal diffusion process in a furnace using BBr_3 as the dopant source at 910 °C, 5 min pre-deposition, and 1000 °C, 7.5 min drive-in step. After the diffusion process, BSG was removed by dipping the wafer in a diluted hydrofluoric acid (HF) solution, followed by DI-water rinsing and drying. The emitter-sheet resistance was changed to 80 ohm/sq using an etch-back method to remove the boron-rich layer (BRL) using an $HF–HNO_3–CH_3COOH$ mixture. The n+ back surface field was formed by diffusing phosphorus with phosphoryl chloride ($POCl_3$) diffusion via 810 °C, 7 min pre-deposition and 880 °C, 5 min drive-in step using the dopant source.

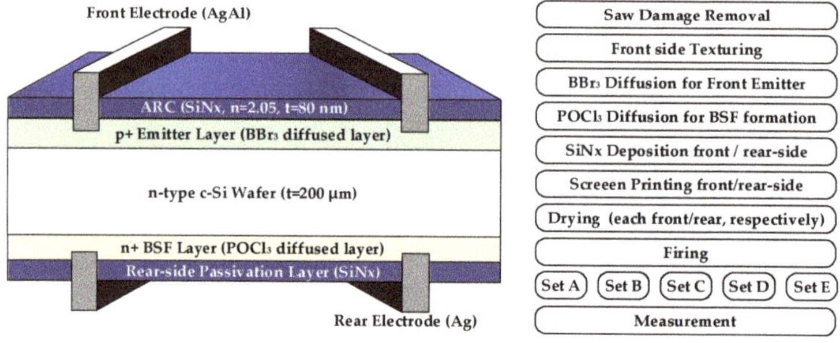

Figure 1. Schematic of the solar cell structure and process flow used in this study.

Then, 80-nm thick SiNx layers with a refractive index of 2.05 were deposited on the front side for both the passivation and anti-reflection coating layers using a plasma-enhanced chemical vapor deposition (PECVD) system with a plasma frequency of 13.56 MHz. In addition, a double-stacked SiNx layer with refractive indices 2.7 and 2.1 was deposited on the rear side for passivation and thermal stability characteristics. Metal electrodes were formed through screen printing using an Ag-Al mixture and Ag paste on the front and rear sides, respectively. After the screen printing process, the samples were dried at 150 °C for a duration of 4 min and subsequently co-fired in a four-zone infrared (IR) belt furnace at the test group temperatures with a peak temperature of 660–930 °C. Twenty-seven cells were produced for each group, and the size of each cell was 10.24 cm^2. The fast-firing profile designs of the experiment were grouped into five different sets, and the detailed characteristics of the co firing groups are listed in Table 1.

Table 1. Fast-firing temperature profile groups of this study.

Co-Firing Temp Group	Temperature (°C)				Peak Temperature (°C)
	Zone 1	Zone 2	Zone 3	Zone 4	
Set A	545	670	570	400	593.4
Set B	545	735	570	400	677.8
Set C	545	800	570	400	719.6
Set D	545	865	570	400	775.0
Set E	545	930	570	400	849.5

For the analysis of the contact phenomenon, the atomic and weight percentages of Ag–Al–Si bonding were analyzed using SEM and EDS. When the fabrication of the solar cell was completed, the cells were characterized. The measurement was performed using illuminated current-voltage (LIV) characteristics (Pasan CT-801, Neuchâtel, Switzerland) under the global solar spectrum of AM 1.5 at 25 °C.

3. Results

Figure 2 shows the variation in R_C with respect to the variation in the peak firing temperature of the fast-firing process. A sample with the same structure as that of the completed cell was fabricated; subsequently, the transmission line (TLM) method pattern was printed, and the Rc was measured. As Figure 2a shows, the co-firing test group A had a very high R_C value, which can be attributed to the low peak firing temperature that was not sufficient to induce a reaction between the metal impurities in the screen-printed paste and the p+ diffused Si surface. Rc dramatically decreased from 181.82 mΩ·cm^2 to 6.98 mΩ·cm^2 for group D at a peak temperature of 865 °C. This revealed that an optimized co-firing temperature is required to initiate a reaction between metal impurities in screen-printing paste and Si atoms.

As the co-firing temperature for Set E was higher than that of Set D, R_C increased from 6.98 mΩ·cm^2 to 10.12 mΩ·cm^2. This increase in R_C may have been due to the variation in the reaction between metal impurities in the screen-printing paste and the Si surface caused by the over-fired electrode. Figure 2b shows the variation in Rc with respect to the variation in the co-firing belt speed. The highest Rc value of 39.19 mΩ·cm^2 was obtained at 60 ipm (inches per min). When the belt speed increased, the R_C value decreased to the lowest value of 6.98 mΩ·cm^2. The excessive co-firing process time effect on the output characteristics of the c-Si solar cell owing to the denaturalization of the metals and other compositions of the electrode. Additionally, the Rc value increased for an increase in belt speed of 110 ipm because of insufficient composition reactions between diffused layer and electrode over the optimal co-firing speed. The actual firing profile for Set D with the lowest Rc was measured using a thermal profiling system (Datapaq Solar Tracker Thermal Profiling System, Fluke, Everett, Washington, DC, USA), and the results are shown in Figure 2b. The measured peak temperature of set D was 775 °C.

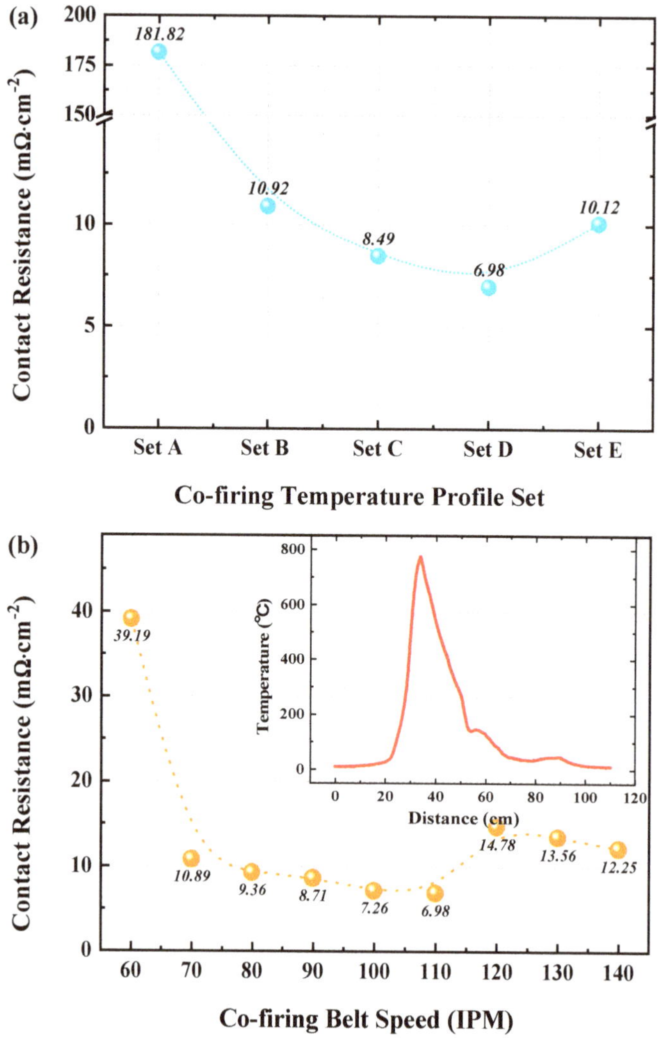

Figure 2. Analysis of R_C of front electrodes formed by the screen-printing process using Ag–Al paste with variation in the co-firing process condition: (**a**) peak temperature and (**b**) belt speed.

For further understanding of the characteristics of the contacts, the metal was etched, and the distribution of the crystallites was studied using SEM (Figure 3). To etch the metal contacts, the cells were dipped in mixed acid, and the glass layer was removed using HF. The removal of the interfacial glass layer made viewing the metal crystallites grown emitter surface easier. For sets A and B, the remaining metal paste was observed, which could be because of the low firing temperature profile. The firing temperature increased, and the color of the metal paste changed. This indicated that the reaction between the metal paste and Si was effective. For set E, the color of the metal paste was darker, which might have been due to over-firing between Si and the metal paste.

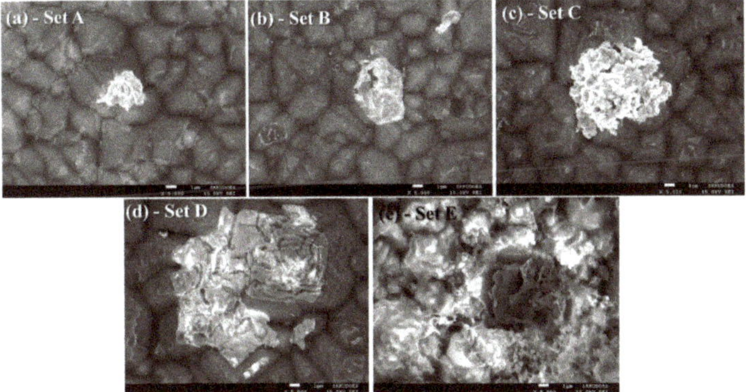

Figure 3. Scanning electron microscopy (SEM) images for the transition state of formed electrode between Si and the front-side electrode when the co-firing peak temperature profile set varied from (**a**) Set A to (**e**) Set E.

The at. wt % of the metals present on the surface were analyzed using EDS. By the electron beam of the SEM, each material emits its own specific X-ray during the sequence in which the electrons in the atom absorb and emit energy. The unique X-ray of each material inside the Ag–Al paste was collected by a detector and classified by intensity to perform a qualitative analysis (atomic/weight percentage) of the front electrode. Figure 4 shows the at. wt % of O, Al, Si, and Ag with variation in the co-firing peak temperature. The EDS results indicated that the Ag content varied dramatically with the co-firing temperature. The at. wt % Si decreased as the co-firing temperature increased, revealing that glass frit and other metal atoms in the metal paste reacted with Si. For set A, as a special scenario, because the co-firing temperature was not sufficient to react with the Si substrate and Ag–Al Paste, most of the paste components disappeared in the HF solution, resulting in a low percentage of Ag and Al and a high proportion of Si. In contrast, the atomic and weight percentages of Ag increased from 16.86% to 34.10% and 46.06% to 67.11%, respectively, as the co-firing temperature increased. Ag atoms were bound to the Si surface after reaction with the glass frit in the metal paste. This determined the binding quality and electrode formation quality between the screen-printed metal paste and the p+ diffused surface of c-Si solar cells.

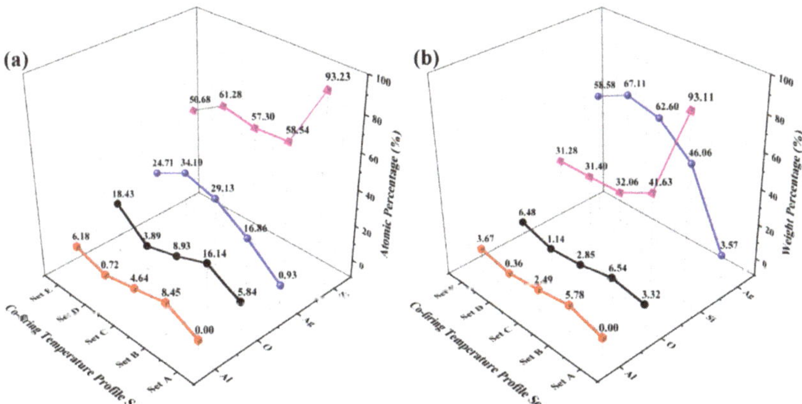

Figure 4. Energy-dispersive X-ray spectroscopy (EDS) analysis of the (**a**) atomic and (**b**) weight percentages of O, Al, Si, and Ag for the co-firing experimental sets A to E.

The at. wt % of Al exhibited a different trend compared with Ag and Si. This indicated that the Al content was bound to the p+ diffused surface with Ag in the Ag–Al metal paste as the co-firing temperature increased. When the co-firing temperature exceeded the optimized process window (>865 °C), Al remained in the electrode and surface because of over firing. The same behavior was observed in the weight percent analysis as the co-firing temperature varied. The difference in the percentage trend of Al in the EDS analysis results will be explained later in this section, but it can be considered as a difference in the contribution of Al to the formation of contact between Ag and Si. For set D, which had the best contact properties, the weight and atomic percentages of Al were the lowest (set A is excluded from this discussion). For set E, which had a higher temperature, Al melted rapidly in the eutectic phase and subsequently cooled, causing Al to escape and form additional voids. A similar pattern was observed in the Al-BSF of the PERC structure [13]. The four-element analysis of the variation in the atomic and weight percentages resulted in contact characteristics between the p+ surface and screen-printed Ag–Al paste determined by the binding ratio of the weight and atomic percentages of Si, Ag, and Al.

Figure 5 shows the SEM images of the alloy formation between the Ag–Al paste and the textured surface for the co-firing profile sets D and E. For the set E sample (Figure 5b), the region between the textured surface and the metal alloy electrode had more voids owing to the over-fired co-firing temperature. A larger void between the pyramid surface and metal electrode decreased the contact characteristics, narrowing down the current path to the electrode. In contrast, for set D, no void region (Figure 5a) occurred between the pyramid surface and metal alloy electrode region. This decreased the current loss.

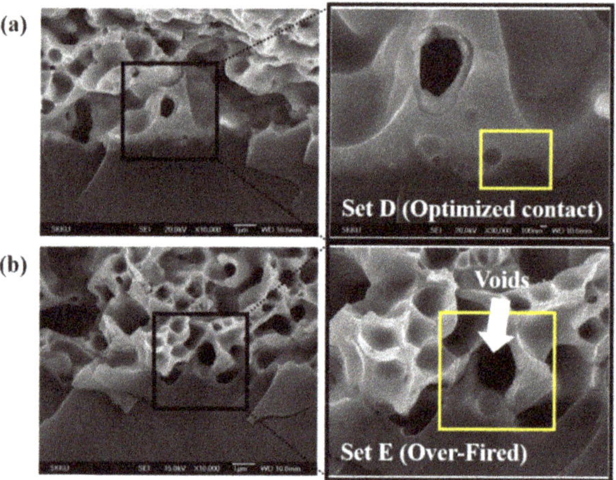

Figure 5. SEM images of the cross-sections of metal electrode formed on a pyramid structure for co-firing temperature profile for (**a**) set D and (**b**) set E.

Figure 6 depicts the reaction diagram of the metal electrode paste and p+ (boron) diffused layer on the Si substrate during the co-firing process after the front and rear-side electrodes were screen-printed. The glass etched the antireflection coating, reduced the melting point of the metal, and promoted adhesion to Si during the firing cycle. The melted glass frits dissolved both Ag and Si. Ag was observed to dissolve in the glass frit when firing occurred. As the glass cooled, the Si regrew epitaxially and Ag crystallites grew into the Si surface from the glass. As described above, the reaction mechanism of the Ag–Al paste had the same pattern until the initial reaction with the general Ag-paste reaction of the p-type substrate excluding Al role in reaction phenomenon. In each firing temperature scenario, Al reacted differently from the p+ diffused and dielectric layers. For Set B, the width of Al and the p+ diffused layer Si alloy (W_{p+}) was shallower than those in the other co-firing profiles because of insufficient reaction energy from thermal temperature. The relationship between W_{p+}, Al, and Si is expressed as [14–16]

$$W_{p+} = \frac{t_{Al} \cdot \rho_{Al}}{\rho_{Si}} \left[\frac{F(T)}{1 - F(T)} - \frac{F(T_o)}{1 - F(T_o)} \right] \quad (2)$$

where t_{Al} is the as-deposited Al thickness, ρ_{Al} and ρ_{Si} are the densities of Al and Si, respectively, $F(T)$ is the at. wt % of the molten phase of Si at the peak alloying temperature, and $F(T_o)$ is the at. wt % of Si at the eutectic temperature (constant). $F(T)$ and $F(T_o)$ were obtained from the Al–Si phase diagram. Thus, the Al–Si atomic percentage determined W_{p+}, and it had a role in the additional current path and locally electrical field similar to that of the Al–BSF layer in conventional p-type Si solar cells. Sets D and E exhibited optimized and over-fired contact phenomena after the co-firing process (Figure 6e). In the overfired scenario, an excessive-W_{p+} layer was formed by a continuous reaction between Al and the p+ diffused layer during the high-temperature co-firing process. In addition, some voids were created at the boundary of the metal paste and the p+ diffused surface by the high reaction temperature and continuous reaction with Ag, Al, Si, and other materials in the electrode metal paste. Consequently, the p–n junction was locally damaged by the excessive-W_{p+} layer, decreasing the device conversion efficiency [17].

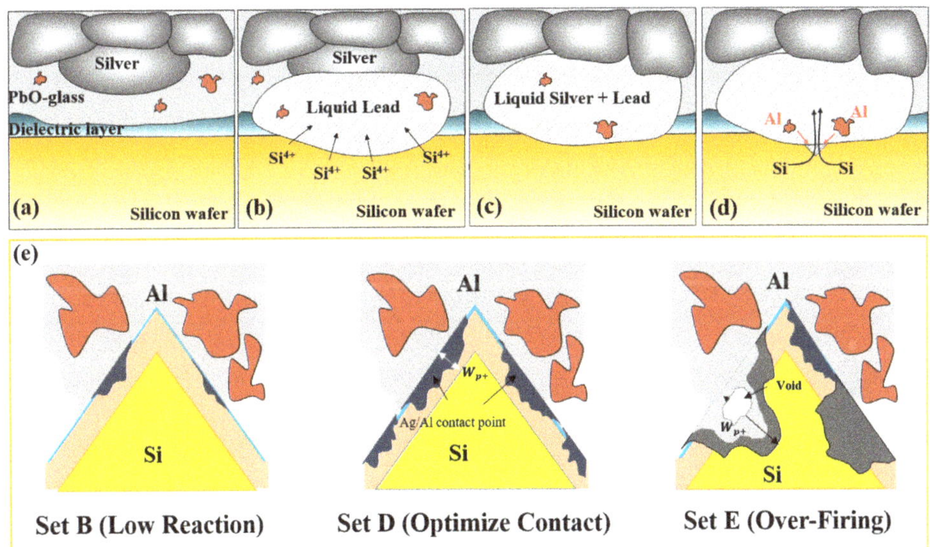

Figure 6. (**a**–**d**) Schematic of the reaction between the Ag–Al paste and the p+ diffused layer as the co-firing process progressed. (**e**) Surface reaction when firing temperature varied for sets B, D, and E.

Figure 7 shows the LIV results of the solar cell fabricated using a co-firing temperature profile set. For set A, the conversion efficiency of the solar cell is not measured because the contact between the front electrode and the p+ diffused layer was not established, as indicated in the SEM/EDS analysis of the electrical properties of the contact. Based on the analysis of the contact properties, the co-firing profile of the other sets were observed to form a sufficient contact characteristic between the p+ diffused layer and the front electrode, unlike that of set A. Although the output characteristics of the solar cell differed, unlike for set A, the conversion efficiency was not measured to be approximately zero. In Set D, which exhibited the best characteristics in terms of Rc, the current density (J_{SC}), open-circuit voltage (V_{OC}), and fill factor (FF) exhibited the best characteristics among the remaining groups, with a conversion efficiency of 19.46%. Moreover, for set E, the Rc value exhibited good properties compared to set B; hence, the fill factor was 60.03%, which was 12% better than that of set B (53.46%). However, owing to the local damage of the p–n junction (as described in Figure 5), an open-circuit voltage characteristic of 471 mV was observed, which was 17% lower than that of set B. The reason why it can be judged that the damage of the p–n junction by Al in the front paste affected the output characteristics of the solar cell, This is because the higher the peak temperature in the contact between the n+ diffused (phosphorus) layer on the back side and the Ag paste, the better it is in terms of contact resistance and fill factor [18].

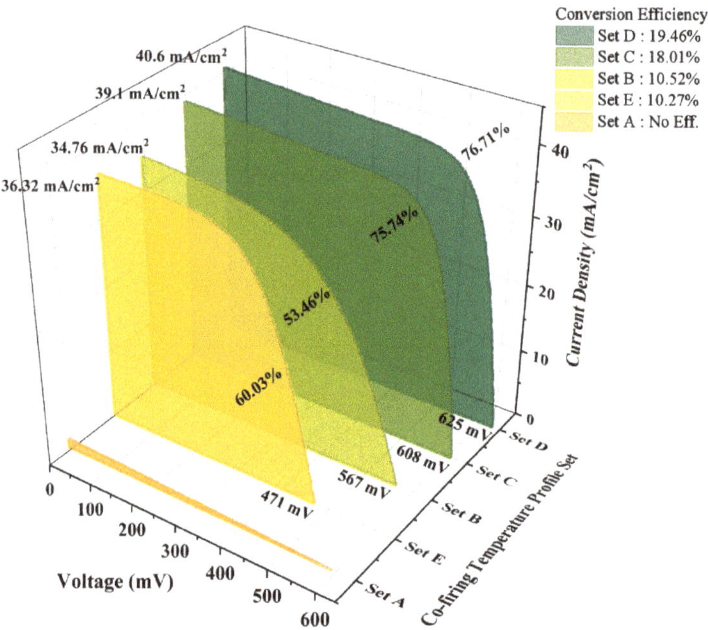

Figure 7. Output illumination current-voltage (LIV) characteristics of the fabricated n-type substrate solar cell for different co-firing temperature profile sets.

Figure 8 shows the loss analysis performed to determine which parameter decreased the output characteristics for the co-firing temperature profile sets B to E (set A was excluded from this analysis owing to low cell output properties in the LIV characteristics). The loss analysis was performed using the software Quokka 3 (March, Germany) technology computer-added design (TCAD) [19–21]. The Rc values obtained from the results shown in Figure 2 were input into the TCAD program. Table 2 lists the detailed TCAD simulation parameters.

In this TCAD simulation analysis, five parameters were selected for the loss analysis: the "front electrode recombination", which applied the characteristics of the front electrode loss such as the finger and busbar region; "lateral carrier transport", which was based on the quality of the interface trap density of the front and rear passivation layer; "current transport loss (resistive)", indicating the contact characteristics between the p+ diffused layer and front electrode; and the "SRH" and "Auger recombination" parameters, related to the recombination properties of the p+ diffused layer.

For set D, the current transport loss had a value of 1.464 mW/cm^2, while the other groups had a value from 1.65 (set C) to 5.52 mW/cm^2 (set B). Because the TCAD simulation was based on the same cell structure, most of the current transport loss could be estimated as the current transfer loss between the p+ diffused layer and the front electrode. In addition, the SRH and Auger recombinations, which represented recombination loss in the p+ diffused layer, exhibited the same loss percentage difference in all groups except set E. The ratio of current transport (resistive) loss accounted for approximately two-thirds of the total energy loss; therefore, it formed the major part of the key loss parameter.

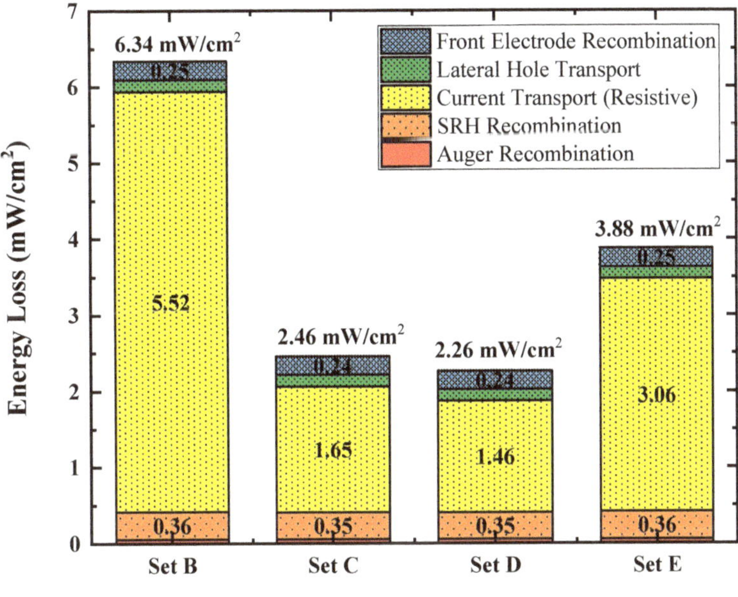

Figure 8. Energy loss analysis of recombination components for co-firing temperature profile sets using a TCAD tool.

Table 2. Detailed TCAD parameters for free energy loss analysis by co-firing temperature profile set.

TCAD Parameter	Set A	Set B	Set C	Set D	Set E
Wafer thickness (nm)			170		
Wafer type			n-type		
Wafer resistivity (Ω cm)			1.0		
Bulk mid-gap SRH lifetimes (μs)			1000 ($\tau_n = \tau_p$)		
p+ diffused emitter R_{sheet} (Ω cm^{-2})			80		
Front j_0: non-contact region(A/cm^2)			30.7×10^{-15}		
Front finger width (μm)			50		
Front contact resistivity (mΩ cm^{-2})	181.82	10.92	8.49	6.98	10.12
Rear local BSF R_{sheet} (Ω cm^{-2})			130		
Rear contact width (μm)			300		
Rear SRV: non-contact region (cm/s)			10		

4. Conclusions

In this study, the variation in the contact properties of the front electrode with the co-firing temperature was studied using SEM/EDS analysis. The proposed profile sets were applied to the production of solar cells to analyze changes in the output characteristics of the solar cells. For the temperature profile of 545, 865, 570, and 400 °C and 110 ipm, the optimal Rc value was 6.98 mΩ-cm^{-2}. This was confirmed by the Al alloy formed between the front electrode and the p+ diffused layer. When the firing temperature was low, the Al alloy did not react sufficiently with the p+ diffused layer. When the firing temperature exceeded the optimum point, it caused damage to the p+ diffused layer and created a void that affected the quality of the p–n junction as well as the current path.

Author Contributions: Conceptualization and writing—original draft preparation, C.P.; Data curation, S.C.; Writing—review and editing, N.B., S.A., S.L. and J.P.; Supervision, J.Y. All authors have read and agreed to the published version of the manuscript.

Funding: This study was supported by the Korea Institute of Energy Technology Evaluation and Planning (KETEP) grant funded by the Korean government (MOTIE) (20203030010310) and by the National Research Foundation of Korea (NRF) grant funded by the Korean government (MSIT) (No. NRF-2019R1A2C1009126).

Conflicts of Interest: The authors declare no conflict of interest.

References

1. *International Technology Roadmap for Photovoltaic (ITRPV)*, 10th ed.; VDMA Photovoltaic Equipment: Frankfurt, Germany, 2019.
2. Lv, Y.; Zhuang, Y.F.; Wang, W.J.; Wei, W.W.; Sheng, J.; Zhang, S.; Shen, W.Z. Towards high-efficiency industrial p-type mono-like Si PERC solar cells. *Sol. Energy Mater. Sol. Cells* **2020**, *204*, 110202. [CrossRef]
3. Yin, H.P.; Tang, K.; Zhang, J.B.; Shan, W.; Huang, X.M.; Shen, X.D. Bifacial n-type silicon solar cells with selective front surface field and rear emitter. *Sol. Energy Mater. Sol. Cells* **2020**, *208*, 110345. [CrossRef]
4. Chen, D.; Chen, Y.; Wang, Z.; Gong, J.; Liu, C.; Zou, Y.; He, Y.; Wang, Y.; Yuan, L.; Lin, W.; et al. 24.58% total area efficiency of screen-printed, large area industrial silicon solar cells with the tunnel oxide passivated contacts (i-TOPCon) design. *Sol. Energy Mater. Sol. Cells* **2020**, *206*, 110258. [CrossRef]
5. Zhang, C.; Shen, H.; Sun, L.; Yang, J.; Wu, S.; Lu, Z. Bifacial p-Type PERC Solar Cell with Efficiency over 22% Using Laser Doped Selective Emitter. *Energies* **2020**, *13*, 1388. [CrossRef]
6. Musztyfaga-Staszuk, M.; Janicki, D.; Panek, P. Correlation of Different Electrical Parameters of Solar Cells with Silver Front Electrodes. *Materials* **2019**, *12*, 366. [CrossRef] [PubMed]
7. Tepner, S.; Wengenmeyr, N.; Ney, L.; Linse, M.; Pospischil, M.; Clement, F. Improving Wall Slip Behavior of Silver Pastes on Screen Emulsions for Fine Line Screen Printing. *Sol. Energy Mater. Sol. Cells* **2019**, *200*, 109969. [CrossRef]
8. Tepner, S.; Ney, L.; Linse, M.; Lorenz, A.; Pospischil, M.; Clement, F. Studying Knotless Screen Patterns for Fine-line Screen Printing of Si-Solar Cells. *IEEE J. Photovolt.* **2020**, *10*, 319–325. [CrossRef]
9. Wagner, H.; Dastgheib-Shirazi, A.; Min, B.; Morishige, A.E.; Steyer, M.; Hahn, G.; del Cañizo, C.; Buonassisi, T.; Altermatt, P.P. Optimizing phosphorus diffusion for photovoltaic applications: Peak doping, inactive phosphorus, gettering, and contact formation. *J. Appl. Phys.* **2016**, *119*, 185704. [CrossRef]
10. Li, H.; Ma, F.-J.; Hameiri, Z.; Wenham, S.; Abbott, M. An Advanced Qualitative Model Regarding the Role of Oxygen During POCl3 Diffusion in Silicon. *Phys. Status Solidi Rapid Res. Lett.* **2017**, *11*, 1700046. [CrossRef]
11. Tsutsui, K.; Matsuda, T.; Watanabe, M.; Jin, C.-G.; Sasaki, Y.; Mizuno, B.; Ikenaga, E.; Kakushima, K.; Ahmet, P.; Maruizumi, T.; et al. Activated boron and its concentration profiles in heavily doped Si studied by soft x-ray photoelectron spectroscopy and Hall measurements. *J. Appl. Phys.* **2008**, *104*, 93709. [CrossRef]
12. Scorzoni, A.; Finetti, M. Metal/Semiconductor contact resistivity and its determination from contact resistance measurements. *Mater. Sci. Rep.* **1988**, *3*, 79–137. [CrossRef]
13. Horbelt, R.; Hahn, G.; Job, R.; Terheiden, B. Void Formation on PERC Solar Cells and Their Impact on the Electrical Cell Parameters Verified by Luminescence and Scanning Acoustic Microscope Measurements. *Energy Procedia* **2015**, *84*, 47–55. [CrossRef]
14. Meemongkolkiat, V.; Hilali, M.; Rohatgi, A. Investigation of RTP and belt fired screen printed AL-BSF on textured and planar back surfaces of silicon solar cells. In Proceedings of the 3rd World Conference on Photovoltaic Energy Conversion, Osaka, Japan, 11–18 May 2003; Volume 1462, pp. 1467–1470.
15. Krause, J.; Woehl, R.; Rauer, M.; Schmiga, C.; Wilde, J.; Biro, D. Microstructural and electrical properties of different-sized aluminum-alloyed contacts and their layer system on silicon surfaces. *Sol. Energy Mater. Sol. Cells* **2011**, *95*, 2151–2160. [CrossRef]
16. Muller, J.; Bothe, K.; Gatz, S.; Plagwitz, H.; Schubert, G.; Brendel, R. Contact Formation and Recombination at Screen-Printed Local Aluminum-Alloyed Silicon Solar Cell Base Contacts. *IEEE Trans. Electron Dev.* **2011**, *58*, 3239–3245. [CrossRef]
17. Urban, T.; Krügel, K.; Heitmann, J. Formation of Ag-Al Alloy in context of PERC solar cell metallization. *Energy Procedia* **2017**, *124*, 930–935. [CrossRef]

18. Rudolph, D.; Olibet, S.; Hoornstra, J.; Weeber, A.; Cabrera, E.; Carr, A.; Koppes, M.; Kopecek, R. Replacement of Silver in Silicon Solar Cell Metallization Pastes Containing a Highly Reactive Glass Frit: Is it Possible? *Energy Procedia* **2013**, *43*, 44–53. [CrossRef]
19. Fell, A.; Schön, J.; Schubert, M.C.; Glunz, S.W. The concept of skins for silicon solar cell modeling. *Sol. Energy Mater. Sol. Cells* **2017**, *173*, 128–133. [CrossRef]
20. Fell, A.; Altermatt, P.P. A Detailed Full-Cell Model of a 2018 Commercial PERC Solar Cell in Quokka3. *IEEE J. Photovolt.* **2018**, *8*, 1443–1448. [CrossRef]
21. Fell, A.; Schon, J.; Muller, M.; Wohrle, N.; Schubert, M.C.; Glunz, S.W. Modeling Edge Recombination in Silicon Solar Cells. *IEEE J. Photovolt.* **2018**, *8*, 428–434. [CrossRef]

Article

Crystallization of Amorphous Silicon via Excimer Laser Annealing and Evaluation of Its Passivation Properties

Sanchari Chowdhury, Jinsu Park, Jaemin Kim, Sehyeon Kim, Youngkuk Kim, Eun-Chel Cho, Younghyun Cho * and Junsin Yi *

College of Information and Communication Engineering, Sungkyunkwan University, Seoul 16419, Korea; sanchari@skku.edu (S.C.); pjsp9@skku.edu (J.P.); rlawoals10@skku.edu (J.K.); kimse429@skku.edu (S.K.); bri3tain@skku.edu (Y.K.); echo0211@skku.edu (E.-C.C.)

* Correspondence: yhcho64@skku.edu (Y.C.); junsin@skku.edu (J.Y.); Tel.: +82-010-5015-8813 (Y.C. & J.Y.)

Received: 22 May 2020; Accepted: 26 June 2020; Published: 30 June 2020

Abstract: The crystallization of hydrogenated amorphous silicon (a-Si:H) is essential for improving solar cell efficiency. In this study, we analyzed the crystallization of a-Si:H via excimer laser annealing (ELA) and compared this process with conventional thermal annealing. ELA prevents thermal damage to the substrate while maintaining the melting point temperature. Here, we used xenon monochloride (XeCl), krypton fluoride (KrF), and deep ultra-violet (UV) lasers with wavelengths of 308, 248, and 266 nm, respectively. Laser energy densities and shot counts were varied during ELA for a-Si:H films between 20 and 80 nm thick. All the samples were subjected to forming gas annealing to eliminate the dangling bonds in the film. The ELA samples were compared with samples subjected to thermal annealing performed at 850–950 °C for a-Si:H films of the same thickness. The crystallinity obtained via deep UV laser annealing was similar to that obtained using conventional thermal annealing. The optimal passivation property was achieved when crystallizing a 20 nm thick a-Si:H layer using the XeCl excimer laser at an energy density of 430 mJ/cm^2. Thus, deep UV laser annealing exhibits potential for the crystallization of a-Si:H films for TOPCon cell fabrication, as compared to conventional thermal annealing.

Keywords: crystallinity; thermal annealing; excimer laser annealing; passivation; amorphous hydrogenated silicon film

---

1. Introduction

Silicon is a formidable material in the electronic and photovoltaic industry due to its outstanding electrical performance and abundancy in nature [1,2]. Polycrystalline silicon (poly-Si) thin films have been widely used in thin film transistors to achieve high definition display in large scale [3]. Also, poly-Si plays important role as a passivating layer in high efficiency crystalline silicon (c-Si) solar cells. The crystallization of amorphous silicon (a-Si) has been achieved using various methods such as thermal annealing, solid-phase crystallization, metal-induced crystallization, sequential lateral solidification, and excimer laser annealing (ELA) [4–7]. The excimer laser is a type of ultra-violet (UV) laser that was first introduced in 1970 by Basov et al. [8]. ELA has been commonly used in the manufacturing of semiconductor integrated circuits. The word "excimer" originates from two words, excimer and dimer, representing a reaction that generates energy [9]. A combination of inert gases, such as Ar, Kr, and Xe, and reactive gases such as F_2 and Cl_2, can be used as the laser gas in ELA. Excimer lasers are advantageous because they are generated under bias and high pressures, which can stop the surface damage caused by high-temperature process. The gas molecules in a closed state are energized, which generates a laser in the UV range. Although Xe and Kr typically do not form compounds when used in excimer lasers, they produce high-energy electrons that yield high-energy pulses in the excitement induced by the beam. These gaseous elements are transient when they are combined with F_2 or Cl_2 gas and form bonded molecules. ELA has been gaining popularity in research as well as in the industry, owing to the short wavelength of an excimer laser, which provides excellent processing quality and high productivity [10]. The crystallization of a-Si using ELA was first achieved in 1994 by the Watanabe group [11,12]. Thereafter, this process has attracted significant research attention because it can be used for heating a thin film of a-Si up to its melting point without causing thermal damage to the substrate. This enables the production of high-quality polysilicon layers. Excimer lasers have outperformed other laser sources due to their larger light source and higher energy density. The UV absorption in excimer lasers is also significantly active and can crystallize a-Si within a comparably short period of time [13].

In this study, we attempted to crystallize a-Si using ELA and evaluated the passivation properties of poly-Si. Our primary motivation was to improve the use of excimer lasers as crystallization sources and enable their application in TOPCon solar cell fabrication technology. Three different excimer laser sources were used as the crystallization sources in this study: xenon monochloride (XeCl) excimer (λ = 308 nm), krypton fluoride (KrF) excimer (λ = 248 nm), and deep UV lasers (λ = 266 nm). The thickness of the a-Si film and the energy density of the laser were varied to analyze the crystallinity of a-Si and evaluate its passivation properties. Past studies showed that crystallinity occurred in lower laser energy density; at higher laser energy density, complete melting of the film occurred [14,15]. Additionally, previous reports showed that a high H content in hydrogenated amorphous silicon (a-Si:H) can be laser crystallized with a remarkable decrease in H content after dehydrogenation process, but for device-grade material, the grain boundary defects of crystallized poly-Si should be passivated with hydrogen [16]. This experimental study targeted optimization of the laser energy density as well as discovery of the best a-Si thickness to improve the effectiveness of ELA in high-efficiency c-Si cell fabrication. We studied the ELA of hydrogenated samples to check their potential as device-grade material. We successfully observed the growth of poly-Si as a function of laser energy density and its effectiveness compared to the conventional thermal annealing process. However, additional research is required to reduce the manufacturing cost and material non-uniformity of the ELA process. This process is also unable to completely crystallize an a-Si layer with a thickness exceeding 50 nm due to its shorter optical penetration depth [17,18]. This paper provides insights into the problems that need to be addressed to ensure that ELA can be effectively and cost-effectively applied on the industrial scale.

2. Materials and Methods

During the experimental procedure, poly-Si formed through the crystallization of a-Si:H achieved via heat treatment and ELA. The samples were prepared using phosphorous doped n-type <100> (3–5 Ω·cm, 200 ± 20 μm thick) Czochralski growth wafers. Surface defects were removed using an etchant solution containing sodium hydroxide (NaOH) powder and sodium hypochlorite (NaOCl) solution at 70–80 °C for 10 min. The samples were then cleaned using the standard Radio Corporation of America (RCA 1 and RCA 2) cleaning methods to remove metallic and organic contaminants from the surface; thereafter, each sample was dipped in hydrochloric acid (HCl) and hydrofluoric acid (HF) for 2 min. Subsequently, the samples were cleaned by rinsing with deionized (DI) water and dried carefully under an N_2 gas atmosphere [19]. The samples were then immediately transferred to the thermal oxide furnace for the deposition of thermally grown silicon dioxide (SiO_2) tunneling layers on both sides of the wafer. An ultra-thin, 1.4 nm tunnel layer SiO_2 was grown via thermal oxidation by exposing the silicon wafers to a temperature of 630 °C under an oxygen gas flow of 5 L/min for 30 min. Thereafter, the a-Si:H layer was deposited on both sides of the wafer using plasma-enhanced chemical vapor deposition (PECVD) with silane (SiH_4) and phosphine (PH_3) gas at a ratio of 1:1. The thickness of the a-Si:H layer was varied from 20 to 80 nm by controlling the flow rate of H_2 gas during PECVD. The samples were then subjected to furnace annealing (FA) (i.e., heat treatment) and

ELA to realize the crystallization of a-Si:H. Heat treatment was performed by varying the temperatures between 850 and 950 °C for 30, 60, and 90 min. When the heat treatment temperature exceeded 900 °C, the tunnel oxide layer was partially damaged due to the high temperature, resulting in a decrease in the passivation effect. Another set of samples was prepared for ELA; these samples were first subjected to dehydrogenation via rapid thermal processing at 450 °C for 60 min to eliminate dangling bonds prior to conducting ELA. The samples were then subjected to ELA using three different lasers: XeCl excimer, KrF excimer, and deep UV lasers at wavelengths of 308, 248, and 266 nm, respectively. The energy density of the XeCl excimer laser was varied from 390 to 450 mJ/cm^2 for a-Si:H film thicknesses of 20, 40, 60, and 80 nm. The energy density of the KrF excimer laser was maintained at 140 mJ/cm^2, while the shot count was changed from 1 to 10 for the same a-Si:H film thickness variation as mentioned for the earlier process. Finally, deep UV laser annealing was performed at an energy density of 660–1980 mJ/cm^2 for a fixed a-Si film thickness of 20 nm and a fixed scan speed of 5 mm/s. The thickness of a-Si:H layer was confirmed from nanoview ellipsometry. Raman spectroscopy for ellipsometry (VASE, JA Woollam, $240 < \lambda < 1700$ nm) was used to measure the crystallinity and passivation by measuring carrier lifetime (τ_{eff}) and implied open circuit voltage (i-V_{OC}) using the WCT Sintrom-120 QSSPC measuring system [20–22]. The experimental results indicated significant progress in crystallization when using ELA compared to that achieved using conventional thermal treatment methods. The heat treatment methods were unable to uniformly crystallize the entire thin film, whereas ELA enabled the selection of a target area for crystallization and the control of the laser energy density [23]. The deep UV ELA samples were subjected to X-ray diffraction (XRD) studies to confirm the crystallinity and increase in grain size (XRD, machine type: Bruker D8 Advance) [24]. The XRD study also confirmed the crystallization of a-Si:H by ELA. These advantages motivated us to employ this experimental approach and find future scopes of ELA for the crystallization of amorphous silicon. A detailed discussion of the experimental results is presented in the following section.

3. Results and Discussions

3.1. Crystallization of a-Si:H Using Heat Treatment

The crystallization of a-Si:H requires a significantly high temperature, which increases the probability of damage to or failure of the sample. Previous studies reported that the minimum temperature required for crystallization of a-Si:H is 600 °C. In this study, crystallization of a-Si:H was conducted at 850–950 °C for 30, 60, and 90 min. Figure 1a presents the Raman shift of the samples subjected to thermal annealing. The sample heated to 900 °C for 30 min exhibited an intense Raman peak. As the thermal annealing temperature increased, the a-Si peak intensity decreased and the poly-Si peak increased and sharpened up to 950 °C. The degree of poly-crystallization and grain growth occurring was clearly observed. However, as the annealing time increased at 950 °C, the agglomeration of poly-Si was observed. A previous study showed that Raman spectra of a-Si:H and poly-Si are located around 480 cm^{-1} and 500–515 cm^{-1}, respectively [25,26]. The Raman spectra in this study confirmed the peak appears around 515–520 cm^{-1}, and the peak intensity increased as the annealing temperature and time increased. Figure 1b shows the crystallinity as a function of annealing temperature and time. The crystallinity increased with the increase in temperature up to 900 °C. A longer annealing time resulted in degradation of crystallinity. Figure 1c depicts the change in τ_{eff} and i-V_{OC} as a function of the heating temperature for the three different treatment times. This figure shows that the optimal passivation property was observed when the samples were treated at 900 °C for 30 min after the forming gas annealing (FGA) treatment, with τ_{eff} of 726 µs and i-V_{OC} of 697 mV. Samples subjected to significantly high temperatures for longer treatment times have a higher probability of undergoing damage to or failure of their passivation layer. A maximum crystallinity of 98.38% was attained when the sample was annealed at 875 °C for 90 min. The degree of crystallinity increased with the heat treatment time and temperature; however, there was a noticeable reduction in the passivation property. This reduction is attributed to the damage in the surface passivation layer caused by exposure to high temperatures for longer treatment times. At such high temperatures, blistering occur and pinholes are created in the passivation layer [27,28].

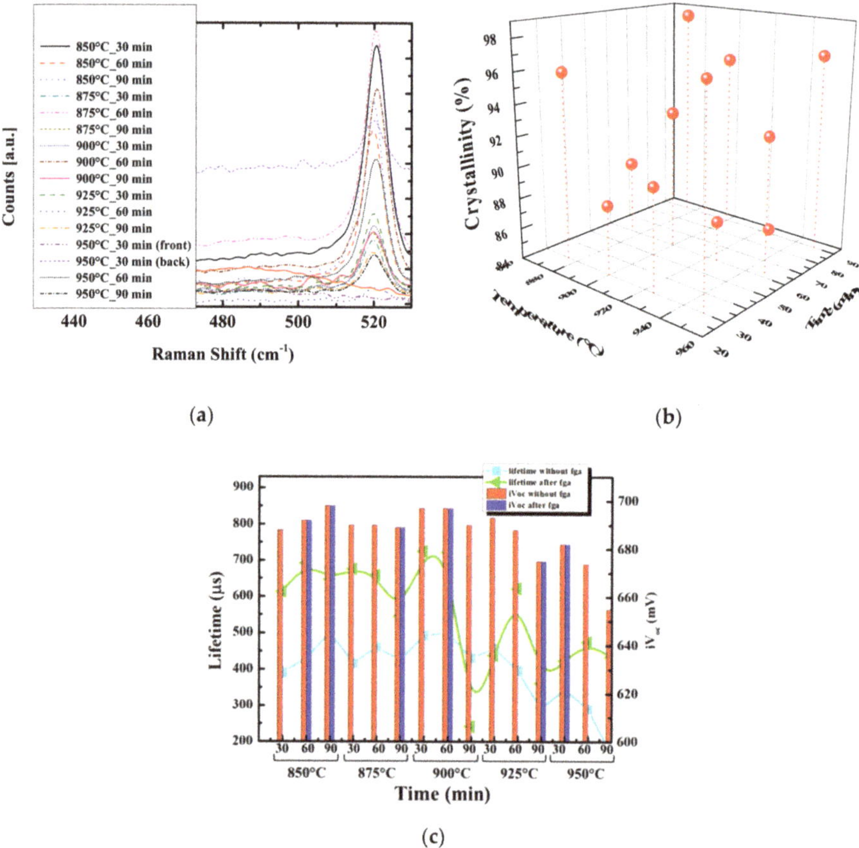

Figure 1. (**a**) Raman spectroscopy; (**b**) change in crystallinity of the a-Si:H thin film after thermal annealing; (**c**) change in the passivation property as a function of the heating temperatures for three different process times.

3.2. Crystallization of a-Si Using Excimer Laser Annealing

Details regarding the a-Si layer crystallization and the evaluation of the passivation property are described in this subsection. As discussed previously, ELA using the deep UV laser was conducted at a laser energy density of 660–1980 mJ/cm^2 with a fixed scan speed of 5 mm/s. The crystallinity and passivation property of the crystallized poly-Si were observed and analyzed. Figure 2a shows the Raman spectra with the increase in laser energy density. As the laser energy density increased, the a-Si peak intensity decreased and the poly-Si peak increased up to 990 mJ/cm^2 energy density. The degree of poly-crystallization and grain growth was clearly observable. The Raman spectra showed that the peak appeared around 515 cm^{-1}, confirming crystallization of poly-Si. The peak intensity reached a maximum at 990 mJ/cm^2. However, further increase in laser energy density resulted in a drastic fall in the peak of poly-Si in Raman shift, indicating the agglomeration of poly-Si. Figure 2b indicates the degree of crystallinity initially increases with an increase in the laser energy density; however, it began to decrease beyond a certain threshold of the energy density. An energy density exceeding 990 mJ/cm^2 resulted in damage to the poly-Si membrane, thereby breaking the lattice structure. The crystallinity also impacted the crystal size, as shown below. Laser energy density also influenced the passivation property of the crystallized a-Si layer. Figure 2c indicates that τ_{eff} and i-V_{OC} significantly increased when the laser energy density increased from 660 to 990 mJ/cm^2; on increasing the energy density further, τ_{eff} and i-V_{OC} decreased considerably. The passivation property was highest when τ_{eff} and i-V_{OC} were 384 µs and 654 mV, respectively, with a crystallinity of approximately 98%. Figure 2d shows the XRD analysis of the samples subjected to deep UV ELA with different laser energy densities. The highest peak intensity was observed from this figure at 990 mJ/cm^2, with a laser energy density around $2\theta = 28.39$, corresponding to Si (111). The figure shows the increase in laser energy density resulted in the decrease in peak intensity, corresponding to Si (111), indicating the degradation in crystallinity. The crystalline size was calculated by the Scherrer equation [29]:

$$D = \frac{K\lambda}{\Delta(2\theta)cos\theta} \tag{1}$$

where θ is the diffraction angle, D is crystalline size, λ is X-ray wavelength, $\Delta(2\theta)$ is the full width half maximum (FWHM), and K is constant. The values of λ and K in this experimental study were 1.54 Å and 0.9, respectively. Table 1 shows the crystalline size of poly silicon film with varying laser energy density along with the crystallinity.

Table 1. The crystalline size and crystallinity of the a-Si:H thin film obtained from deep UV laser annealing for different laser energy density.

Laser Energy Density (mJ/cm^2)	660	990	1320	1650	1980
Crystalline Size (D, nm)	86	140	75	78	66
Crystallinity (X_c, %)	64	97.9	58.1	55.6	34.4

The surface morphology of the samples subjected to deep UV laser annealing was also observed from field emission scanning electron microscopy (FESEM) images. The images shown in Figure 3a,b indicate that the grain size of globular crystallites increased with an increase in energy density up to 990 mJ/cm^2; in Figure 3c, the broken lattice structure shown was caused by the energy density exceeding 990 mJ/cm^2. These FESEM images confirmed that higher laser energy density results in a broken lattice structure, resulting in degradation in crystallinity.

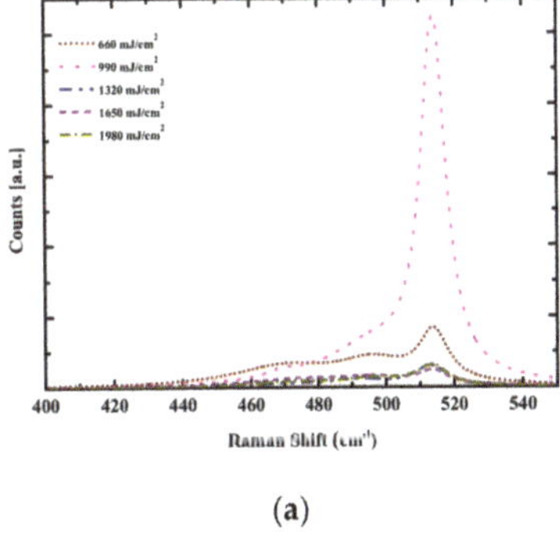

(a)

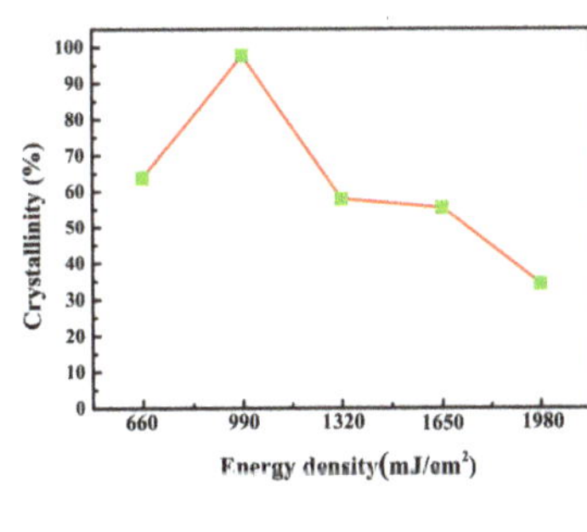

(b)

Figure 2. *Cont.*

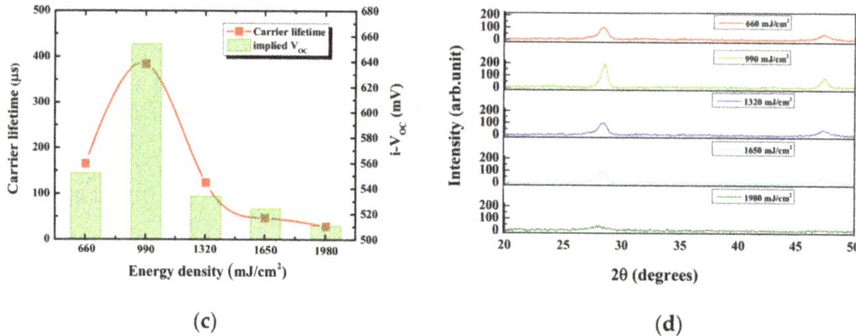

(c) (d)

Figure 2. (**a**) Raman spectroscopy; (**b**) change in crystallinity of the a-Si:H thin film after deep UV ELA for various laser energy densities; (**c**) change in the passivation property as a function of laser energy density for the crystallization of a 20 nm thick a-Si:H layer; (**d**) XRD of the crystallize a-Si:H thin film subjected to deep UV laser annealing for different energy density.

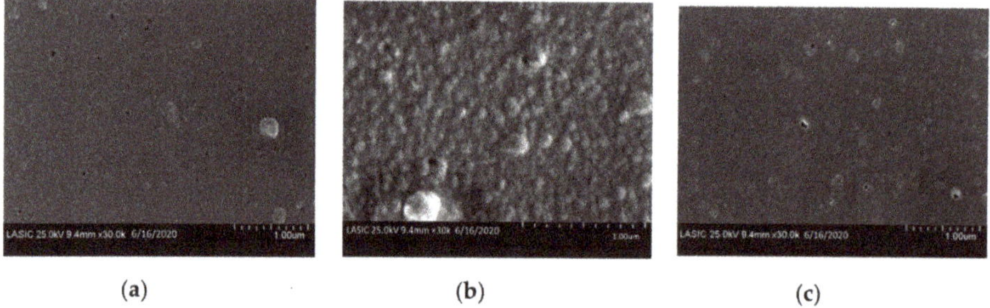

(a) (b) (c)

Figure 3. SEM images of the a-Si:H films after crystallization via deep UV ELA at energy densities of (**a**) 660, (**b**) 990, and (**c**) 1320 mJ/cm^2.

During the crystallization of a-Si using the XeCl excimer laser, the energy density was varied from 390 to 450 mJ/cm^2, with a varied a-Si layer thickness of 20–80 nm. The samples were cured at 900 °C for 60 min to achieve higher oxidation and better passivation. Figure 4 presents the passivation property observed under these conditions. This figure indicates that the oxidation during ELA increased as the a-Si layer thickness decreased and the laser energy density increased; this occurred because the optical penetration depth for ELA is shorter, making it difficult to crystallize thick a-Si layers. When the laser energy density exceeded 430 mJ/cm^2, the passivation property decreased due to damage to the a-Si layer.

The crystallization of a-Si:H using KrF excimer laser was performed. The laser energy density was kept constant at 140 mJ/cm^2 while varying the laser shot count from 1 to 10 for a-Si layer thicknesses of 20–80 nm. The samples were subjected to curing at 900 °C prior to ELA, and FGA was conducted subsequently. This significantly improved passivation. Figure 5a,b indicate the variation in the passivation property as a function of the a-Si layer thickness for laser shot counts of 1 to 10. This figure shows that the passivation property increased with the laser shot count in the case of a thin a-Si:H layer; however, an increase in the thickness of the a-Si layer significantly reduced the passivation property. Highest τ_{eff} and i-V_{OC} values of 353 μs and 674 mV, respectively, were obtained after the FGA process as shown in Figure 5a. However, the experimental results also indicated that the ELA process conditions need to be further optimized to reduce manufacturing costs and enable the large-scale application of this technique for efficient production.

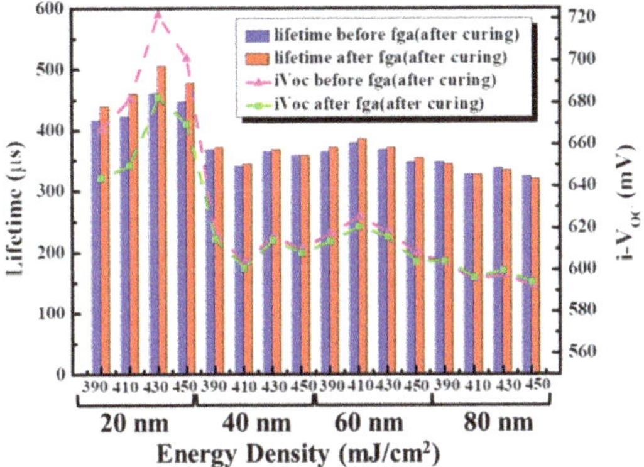

Figure 4. Variation in the passivation property as a function of the energy density of the XeCl excimer laser for four different a-Si:H layer thicknesses.

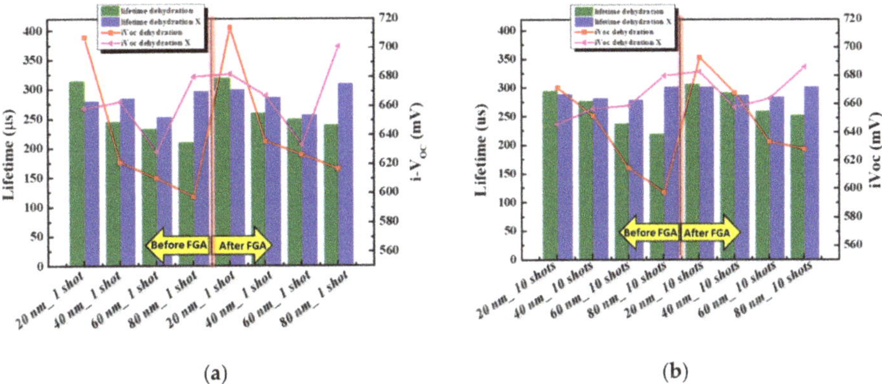

Figure 5. Variation in the passivation property of the a-Si:H film after crystallization via KrF ELA for (**a**) 1 and (**b**) 10 shots count.

4. Conclusions

This experimental study focused on the crystallization of an a-Si:H layer using ELA instead of the conventional thermal annealing to prevent damage to the sample caused by the high-temperature process. The potential of ELA to become a key method for crystallization in the photovoltaic and opto-electronic device market was demonstrated in this paper. The sample subjected to deep UV laser annealing at an energy density of 990 mJ/cm^2 exhibited a high degree of crystallinity. This indicates the potential applicability of ELA for crystallization compared to the conventional heat treatment processes. However, the value of τ_{eff} achieved via heat-treatment crystallization was 727 μs, which is higher than that achieved using ELA (588 μs). This outcome suggests that the ELA process requires further optimization for achieving high passivation properties in the crystallized a-Si:H layer. Hence, additional research is required to control the laser energy density, laser shot counts, and process time, with the aim of improving passivation properties of the crystallized layer. The XeCl and KrF excimers lasers also yielded improved crystallinity in thin a-Si layers; however, their performances degraded when the thickness of the layer increased. Hence, these excimer lasers are not suitable because a-Si:H layers used as passivation layers in c-Si solar cells are required to have higher thicknesses. This research provides valuable insights into the necessity of optimizing the ELA conditions, with the aim of increasing the optical penetration depth and controlling the energy density. This technology should be cost effective to ensure its wide-spread application for manufacturing high-efficiency solar cells.

Author Contributions: Writing—original draft S.C.; Data Curation—S.C., S.K., J.P., J.K., and Y.K.; Formal Analysis—E.-C.C., G.T.C.; Project—Administration, Y.K., E.-C.C., Y.C., and J.Y.; Supervision—J.Y.; Writing—review & editing—G.T.C. and E.-C.C. All authors have read and agreed to the published version of the manuscript.

Funding: This Research is funded by the Korea Institute of Energy Technology Evaluation and Planning grant funded by the Korea Government.

Acknowledgments: This work was supported by the Korea Institute of Energy Technology Evaluation and Planning (KETEP) and the Ministry of Trade, Industry and Energy (MOTIE) of the Republic of Korea (No. 20194010000090 and 20183010014270).

Conflicts of Interest: The authors declare no conflict of interest.

References

1. Dassow, R.; Köhler, J.R.; Grauvogl, M.; Bergmann, R.B.; Werner, J. Laser-Crystallized Polycrystalline Silicon on Glass for Photovoltaic Applications. *Solid State Phenom.* **1999**, *67*, 193–198. [CrossRef]
2. Staebler, D.L.; Wronski, C.R. Reversible conductivity changes in discharge-produced amorphous Si. *Appl. Phys. Lett.* **1977**, *31*, 292–294. [CrossRef]
3. Mei, P.; Boyce, J.B.; Hack, M.; Lujan, R.; Ready, S.E.; Fork, D.K.; Johnson, R.I.; Anderson, G.B. Grain growth in laser dehydrogenated and crystallized polycrystalline silicon for thin film transistors. *J. Appl. Phys.* **1994**, *76*, 3194–3199. [CrossRef]
4. Im, J.S.; Kim, H.J.; Thompson, M.O. Phase transformation mechanisms involved in excimer laser crystallization of amorphous silicon films. *Appl. Phys. Lett.* **1993**, *63*, 1969–1971. [CrossRef]
5. Miyasaka, M.; Stoemenos, J. Excimer laser annealing of amorphous and solid-phase-crystallized silicon films. *J. Appl. Phys.* **1999**, *86*, 5556–5565. [CrossRef]
6. Smith, P.M.; Carey, P.G.; Sigmon, T.W. Excimer laser crystallization and doping of silicon films on plastic substrates. *Appl. Phys. Lett.* **1997**, *70*, 342–344. [CrossRef]
7. Song, D.; Inns, D.; Straub, A.; Terry, M.L.; Campbell, P.; Aberle, A.G. Solid phase crystallized polycrystalline thin-films on glass from evaporated silicon for photovoltaic applications. *Thin Solid Film.* **2006**, *513*, 356–363. [CrossRef]
8. Brotherton, S.D.; McCulloch, D.J.; Clegg, J.B.; Gowers, J.P. Excimer-laser-annealed poly-Si thin-film transistors. *IEEE Trans. Electron Devices* **1993**, *40*, 407–413. [CrossRef]
9. Komotori, J.; Lee, B.J.; Dong, H.; Dearnley, P.A. Corrosion response of surface engineered titanium alloys damaged by prior abrasion. *Wear* **2001**, *251*, 1239–1249. [CrossRef]
10. Sato, H.; Nishio, S. Polymer laser photochemistry, ablation, reconstruction, and polymerization. *J. Photochem. Photobiol. C Photochem. Rev.* **2001**, *2*, 139–152. [CrossRef]
11. Love, R.A.; Parge, H.E.; Wickersham, J.A.; Hostomsky, Z.; Habuka, N.; Moomaw, E.W.; Adachi, T.; Hostomska, Z.J.C. The crystal structure of hepatitis C virus NS3 proteinase reveals a trypsin-like fold and a structural zinc binding site. *Cell* **1996**, *87*, 331–342. [CrossRef]
12. Mohan, S. *Fiber Optics and Laser Instrumentation: (for EEE, EI, Electronics, Computer Science & Engineering, Physics and Materials Science Students in Indian Universities)*; MJP Publisher: Tamil Nadu, India, 2019.
13. Carluccio, R.; Cina, S.; Fortunato, G.; Friligkos, S.; Stoemenos, J. Structure of poly-Si films obtained by laser annealing. *Thin Solid Film.* **1997**, *296*, 57–60. [CrossRef]
14. Kuriyama, H.; Nohda, T.; Ishida, S.; Kuwahara, T.; Noguchi, S.; Kiyama, S.; Tsuda, S.; Nakano, S. Lateral Grain Growth of Poly-Si Films with a Specific Orientation by an Excimer Laser Annealing Method. *Jpn. J. Appl. Phys.* **1993**, *32*, 6190–6195. [CrossRef]
15. McCulloch, D.J.; Brotherton, S.D. Surface roughness effects in laser crystallized polycrystalline silicon. *Appl. Phys. Lett.* **1995**, *66*, 2060–2062. [CrossRef]
16. Boyce, J.B.; Mei, P. Laser Crystallization for Polycrystalline Silicon Device Applications. In *Technology and Applications of Amorphous Silicon*; Street, R.A., Ed.; Springer: Berlin/Heidelberg, Germany, 2000; pp. 94–146.
17. Bidin, N.; Ab Razak, S.N. ArF Excimer Laser Annealing of Polycrystalline Silicon Thin Film. In *Crystallization: Science and Technology*; Intech Open Publisher: London, UK, 2012; p. 481.

18. Kühnapfel, S.; Severin, S.; Kersten, N.; Harten, P.; Stegemann, B.; Gall, S. Multi crystalline silicon thin films grown directly on low cost soda-lime glass substrates. *Sol. Energy Mater. Sol. Cells* **2019**, *203*, 110168. [CrossRef]
19. Mallem, K.; Kim, Y.J.; Hussain, S.Q.; Dutta, S.; Le, A.H.T.; Ju, M.; Park, J.; Cho, Y.H.; Kim, Y.; Cho, E.-C.; et al. Molybdenum oxide: A superior hole extraction layer for replacing p-type hydrogenated amorphous silicon with high efficiency heterojunction Si solar cells. *Mater. Res. Bull.* **2019**, *110*, 90–96. [CrossRef]
20. Kimmerle, A.; Rothhardt, P.; Wolf, A.; Sinton, R.A. Increased Reliability for J0-analysis by QSSPC. *Energy Procedia* **2014**, *55*, 101–106. [CrossRef]
21. Koo, K.; Kim, S.; Choi, P.; Kim, J.; Jang, K.; Choi, B. Electrical evaluation of the crystallization characteristics of excimer laser annealed polycrystalline silicon active layer. *Jpn. J. Appl. Phys.* **2018**, *57*, 106503. [CrossRef]
22. Mall, A.K.; Paul, B.; Garg, A.; Gupta, R. Temperature dependent X-ray diffraction and Raman spectroscopy studies of polycrystalline $YCrO_3$ ceramics across the TC ~460 K. *J. Raman Spectrosc.* **2020**, *51*, 537–545. [CrossRef]
23. Pangal, K.; Sturm, J.C.; Wagner, S.; Büyüklimanli, T.H. Hydrogen plasma enhanced crystallization of hydrogenated amorphous silicon films. *J. Appl. Phys.* **1999**, *85*, 1900–1906. [CrossRef]
24. Meador, J.A. *A Radiological Survey of Two Bruker D8 Advance XRD Instruments Located in Separate Quality Control Laboratories*; West Virginia University Research Repository: Morgantown, WV, USA, 2016.
25. Saleh, R.; Nickel, N.H. Raman spectroscopy of doped and compensated laser crystallized polycrystalline silicon thin films. *Surf. Coat. Technol.* **2005**, *198*, 143–147. [CrossRef]
26. Yoo, W.S.; Ishigaki, T.; Ueda, T.; Kang, K.; Kwak, N.Y.; Sheen, D.S.; Kim, S.S.; Ko, M.S.; Shin, W.S.; Lee, B.S.; et al. Grain size monitoring of 3D flash memory channel poly-Si using multiwavelength Raman spectroscopy. In Proceedings of 2014 14th Annual Non-Volatile Memory Technology Symposium (NVMTS), Jeju Island, Korea, 27–29 October 2014; pp. 1–4.
27. Morisset, A.; Cabal, R.; Grange, B.; Marchat, C.; Alvarez, J.; Gueunier-Farret, M.-E.; Dubois, S.; Kleider, J.-P. Highly passivating and blister-free hole selective poly-silicon based contact for large area crystalline silicon solar cells. *Sol. Energy Mater. Sol. Cells* **2019**, *200*, 109912. [CrossRef]
28. Spruiell, J. *A Review of the Measurement and Development of Crystallinity and Its Relation to Properties in Neat Poly (Phenylene Sulfide) and Its Fiber Reinforced Composites*; Oak Ridge National Lab. (ORNL): Oak Ridge, TN, USA, 2005.
29. Monshi, A.; Foroughi, M.R.; Monshi, M.R. Modified Scherrer equation to estimate more accurately nano-crystallite size using XRD. *World J. Nano Sci. Eng.* **2012**, *2*, 154–160. [CrossRef]

Article

Influence of the Carrier Selective Front Contact Layer and Defect State of a-Si:H/c-Si Interface on the Rear Emitter Silicon Heterojunction Solar Cells

Sunhwa Lee [1], Duy Phong Pham [1], Youngkuk Kim [1], Eun-Chel Cho [1], Jinjoo Park [2,*] and Junsin Yi [1,*]

1 School of Information and Communication Engineering, Sungkyunkwan University, Suwon 16419, Korea; ssunholic@skku.edu (S.L.); pdphong@skku.edu (D.P.P.); bri3tain@skku.edu (Y.K.); echo0211@skku.edu (E.-C.C.)

2 Division of Energy and Optical Technology Convergence, Cheongju University, Cheongju 28503, Korea

* Correspondence: jwjh3516@cju.ac.kr (J.P.); junsin@skku.edu (J.Y.)

Received: 30 April 2020; Accepted: 4 June 2020; Published: 8 June 2020

Abstract: In this research, simulations were performed to investigate the effects of carrier selective front contact (CSFC) layer and defect state of hydrogenated amorphous silicon passivation layer/n-type crystalline silicon interface in silicon heterojunction (SHJ) solar cells employing the Automat for Simulation of hetero-structure (AFORS-HET) simulation program. The results demonstrated the effects of band offset determined by band bending at the interface of the CSFC layer/passivation layer. In addition, the nc-SiOx: H CSFC layer not only reduces parasitic absorption loss but also has a tunneling effect and field effect passivation. Furthermore, it increased the selectivity of contact. In the experimental cell, nc-SiO_x:H was used as the CSFC layer, where efficiency of the SHJ solar cell was 22.77%. Our investigation shows that if a SiO_x layer passivation layer is used, the device can achieve efficiency up to 25.26%. This improvement in the cell is mainly due to the enhancement in open circuit voltage (V_{oc}) because of lower interface defect density resulting from the SiO_x passivation layer.

Keywords: carrier selective contact; rear emitter heterojunction; passivation

---

1. Introduction

Crystalline silicon (c-Si) solar cells dominate the global photovoltaic market, accounting for more than 90% of production [1,2]. Persistent efforts have been undertaken to achieve their theoretical efficiency threshold of 29.4% by improving the fundamental limiting factors. Recently, an efficiency level of 26.7% for a silicon heterojunction (SHJ) cell was reported by Kaneka corporation using the integrated back contact technique [3]. This result shows the potential of SHJ cell technology to achieve extremely high efficiency. Hence, this technology is expected to not only improve cell efficiency but also reduce manufacturing cost. The performance improvement of SHJ solar cells is dependent on reduction in the following: 1) carrier collection losses using thin wafer [4]; 2) surface recombination losses by surface passivation [5,6]; and 3) parasitic absorption loss of carrier selective front contact (CSFC) layers by controlling their thickness and using wide band-gap materials [7,8]. Interface defects and the optical parasitic absorption loss of the CSFC layers remain the key restrictions on the efficiency of SHJ solar cells.

Conventional SHJ solar cells are generally configured with front p-type hydrogenated amorphous silicon (p-a-Si:H) contact layers to collect better minority carriers [9]. However, due to high activation energy and the low band-gap of such front layers, a Schottky barrier is formed between the transparent conductive oxide (TCO) and p-a-Si:H layers [10,11]. Heavily doped p-layers can be used to reduce the Schottky barrier, but this causes parasitic absorption and increased carrier recombination. Hence, a rear emitter SHJ solar cell (RE-SHJ) with a p-layer located on the rear side has been analyzed [12]. Conventional RE-SHJ solar cells use doped a-Si:H material as both the front and back carrier selective contact layers. The internal electric field is formed according to the doping concentration of these doped layers and, as a result, the carriers are separated and collected. As the doping concentration increases, doping-induced defects also increase, which limits the efficiency of the device. Additionally, the a-Si:H materials reduce transmittance owing to the inherent low band-gap. A wide-gap hydrogenated nanocrystalline silicon oxide (nc-SiO_X:H) layer, which compensates for the disadvantages of a-Si:H [13], is a suitable alternative for the efficiency improvement of RE-SHJ devices. Compared with a-Si:H, nc-SiO_X:H has a low absorption coefficient in the short wavelength region, as well as excellent conduction properties owing to the crystalline phase and quantum confinement effects [14]. In this paper, we present the numerical simulation of RE-SHJ cells using Automat for simulation hetero-structures (AFORS-HET) software. The influence of a-Si:H and silicon oxide materials functionalized as passivation and CSFC layers for the RE-SHJ cells is discussed under experimental and simulation results. The simulation is performed to identify the essential factors for enhancing the efficiency of the device. A numerical simulation is performed in conjunction with experimental data (absorption coefficient, extinction coefficient and optical band gap for each layer) to evaluate the conditions required to increase the efficiency of the solar cells.

2. Materials and Methods

2.1. Experimental Methods

In the device fabrication, a Czochralski (CZ)-grown n-type silicon wafer (148 μm thickness and 1.5 Ωcm resistivity) was used. The surface texturing was performed using sodium hydroxide (8% NaOH) solution. Radio Corporation of America (RCA) standard cleaning was used to remove the surface impurities from the texture wafers and was applied in two stages: standard cleaning 1 and 2. Subsequently, a thin a-Si:H passivation layer was prepared on both interfaces of the Si-wafer. The doped n-type layer was deposited at the front side, and the p-type layer was deposited at the rear. These layers were deposited by employing a standard plasma-enhanced chemical vapor deposition system. The front and back TCO layers were deposited using the radio frequency (RF) sputtering system, which requires Ar feed gas and approximately 200 °C substrate heating during deposition. Additionally, screen printing with Ag paste was used for the front metal grid and back contact electrodes. The RE-SHJ cell structure consisted of front electrode/TCO/n-type CSFC layer/a-Si:H passivation layer/c-Si wafer/a-Si:H passivation layer/p-layer emitter/TCO/back electrode, as shown in Figure 1.

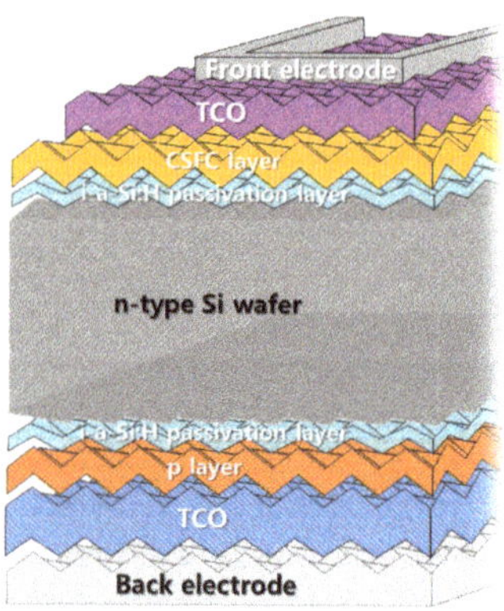

Figure 1. Schematic structure of a heterojunction solar cell.

For the wafer surface passivation with SiO_x layers, the SiO_x passivation layer was formed by varying the pressure of CO_2 plasma post-treatment from 1600 to 2200 mTorr on the a-Si/Si-wafer interface. The minority carrier lifetime (τ_{eff}) was examined using quasi-steady-state photo conductance (QSSPC, WCT-120). For measuring spectroscopic ellipsometry (SE) and X-ray photoelectron spectroscopy, a <100> n-type CZ Si wafer (1–10 Ωcm) with one-side polished was used. The thickness, refractive index, absorption coefficient and optical bandgap were measured by SE (VASE, J. A. Woollam) in the 240 to 1700 nm wavelength range. An X-ray photoelectron spectroscopy (XPS, K-Alpha, Thermo Scientific) system was utilized to estimate the chemical composition using Al Kα radiation.

2.2. Modeling of HIT Solar Cells

The recombination in the semiconductor was calculated using Shockley–Read–Hall statistics [15]. The defect state was investigated at the passivation layer/Si-wafer interface by varying the defect density in the range of 1×10^{10} to 1×10^{14} $cm^{-2}eV^{-1}$. The illuminated current–voltage characteristics were measured at AM 1.5G (100 mW/cm^2). The Gaussian defect distribution was utilized to describe the band tail states of a-Si:H and nc-SiO_x:H layers in both valence and conduction band. Due to the n-type c-Si substrate, oxide energy levels with an average defect density of 1×10^{11} $cm^{-2}eV^{-1}$ were settled at 0.53 eV from the conduction band. The simulated parameters are listed in Table 1 for all layers. The RE-SHJ solar cell with n-a-Si:H CSFC layer (Figure 2a) consisted of front electrode/TCO/n-a-Si:H CSFC layer/i-a-Si:H passivation layer/n-type c-Si wafer/i-a-Si:H passivation layer/p-a-Si:H layer/TCO/back electrode. The RE-SHJ solar cell with nc-a-SiO_x:H CSFC layer (Figure 2b) consisted of front electrode/TCO/n-a-Si:H CSFC layer/i-a-Si:H passivation layer/n-type c-Si wafer/i-a-Si:H passivation layer/p-nc-SiO_x:H layer/TCO/back electrode.

Table 1. Material parameter values for SHJ solar cell simulation.

Parameters	n-a-Si:H	n-nc-SiOx:H	i-a-Si:H	p-a-Si:H	p-nc-SiOx:H	n-c-Si
Thickness (nm)	5	20	3	5	5	1.5×10^5
Dielectric constant	11.9	11.9	11.9	11.9	11.9	11.9
Electron affinity (eV)	3.9	3.95	3.9	3.9	3.9	4.05
Bandgap (eV)	1.72	2.1	1.70	1.72	2.05	1.124
Effective conduction band density (cm^{-3})	1×10^{21}	1×10^{20}	1×10^{20}	1×10^{20}	1×10^{20}	2.8×10^{19}
Effective valence band density (cm^{-3})	1×10^{21}	1×10^{20}	1×10^{20}	1×10^{20}	1×10^{20}	2.2×10^{19}
Effective electron mobility (cm^2/Vs)	5	50	5	5	20	858
Effective hole mobility (cm^2/Vs)	1	5	1	1	5	355
Doping concentration acceptors (cm^{-3})	0	0	0	1×10^{18}	1×10^{18}	0
Doping concentration donators (cm^{-3})	1×10^{18}	1.42×10^{19}	0	0	0	5.63×10^{16}
Electron thermal velocity (cm/s)	1×10^7	1×10^7	1×10^7	1×10^7	1×10^7	1×10^7
Hole thermal velocity (cm/s)	1×10^7	1×10^7	1×10^7	1×10^7	1×10^7	1×10^7
Parameters	**n-a-Si:H**	**n-nc-SiOx:H**	**i-a-Si:H**	**p a-Si:H**	**p-nc-SiOx:H**	
Defect density at conduction band (CB) edge ($cm^{-3}eV^{-1}$)	1×10^{21}	2×10^{19}	1×10^{21}	1×10^{21}	1×10^{21}	
Defect density at valence band (VB) edge ($cm^{-3}eV^{-1}$)	1×10^{21}	2×10^{19}	1×10^{21}	1×10^{21}	1×10^{21}	
Urbach energy for CB tail (eV)	0.025	0.025	0.025	0.035	0.035	
Urbach energy for VB tail (eV)	0.03	0.03	0.025	0.025	0.025	
$\sigma_e(\sigma_h)$ for CB tail (cm^{-2})	7×10^{-16} (7×10^{-16})	1×10^{-17} (1×10^{-15})	7×10^{-16} (7×10^{-16})	7×10^{-16} (7×10^{-16})	7×10^{-16} (7×10^{-16})	
$\sigma_e(\sigma_h)$ for VB tail (cm^{-2})	7×10^{-16} (7×10^{-16})	1×10^{-15} (1×10^{-17})	7×10^{-16} (7×10^{-16})	7×10^{-16} (7×10^{-16})	7×10^{-16} (7×10^{-16})	
Donor (Acceptor) like Gaussian density of states ($cm^{-3}eV^{-1}$)	2.5×10^{-17} (2.5×10^{-17})	1×10^{-17} (1×10^{-17})	2.5×10^{-17} (2.5×10^{-17})	2.5×10^{-17} (2.5×10^{-17})	1.1×10^{-17} (1.1×10^{-17})	
Gaussian peak energy for donor (eV)	0.45	0.46	0.7	1.02	1.02	
Gaussian peak energy for acceptor (eV)	0.65	0.65	1	1.2	1.2	
$\sigma_e(\sigma_h)$ for acceptor like Gaussian states (cm^{-2})	3×10^{-15} (3×10^{-14})	3×10^{-15} (3×10^{-14})	3×10^{-15} (3×10^{-14})	3×10^{-15} (3×10^{-14})	3×10^{-15} (3×10^{-14})	
$\sigma_e(\sigma_h)$ for donor like Gaussian states (cm^{-2})	3×10^{-14} (3×10^{-15})	3×10^{-14} (3×10^{-15})	3×10^{-14} (3×10^{-15})	3×10^{-14} (3×10^{-15})	3×10^{-14} (3×10^{-15})	

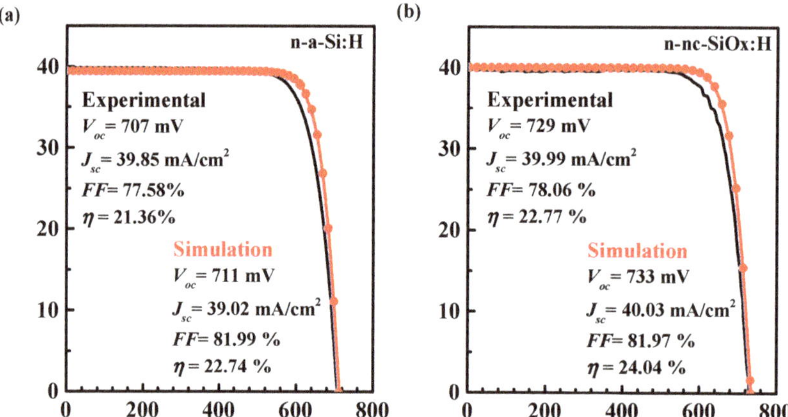

Figure 2. J-V curve characteristics of rear emitter silicon heterojunction (SHJ) solar cells with different carrier selective front contact (CSFC) layers: (**a**) a-Si:H; (**b**) a-SiO_x:H. Experimental results are indicated by the continuous line, while the simulation results are indicated by the symbol.

3. Results and Discussion

3.1. Numerical Simulation of HIT Solar Cell

The SHJ solar cells use a-Si:H materials as the passivation and carrier selective contact layers at both sides for separate electron-hole pairs. The conventional SHJ solar cell uses the p-a-Si:H layer as a CSFC layer. This layer absorbs light in the short wavelength region and has low conductivity, which leads to parasitic absorption losses. The simulation of the rear emitter with various CSFC layers and interface defect states was investigated. The simulations were performed using optical parameters of layers, including extinction coefficient (k), refractive index (n), and optical bandgap (E_g) obtained via ellipsometry, as well as electrical parameter measurements. The J-V curve characteristics of the RE-SHJ cells with different CSFC layers were compared to that of real cells, as shown in Figure 3. It can be observed that the characteristics of the simulated results are very similar to that of the real cell; for example, while fill factor (*FF*) depicted a higher deviation, the other parameters were very close to those of the real cells. The nc-SiO_X:H CSFC layer achieved a conversion efficiency of 24.04%. The decrease in parasitic absorption of the wide-gap nc-SiO_X:H CSFC layer assisted the increase in short circuit current density (J_{sc}). The increase in efficiency of the RE-SHJ cell using the nc-SiO_X:H CSFC layer is explained through the band diagram, electric field and generation/recombination rate.

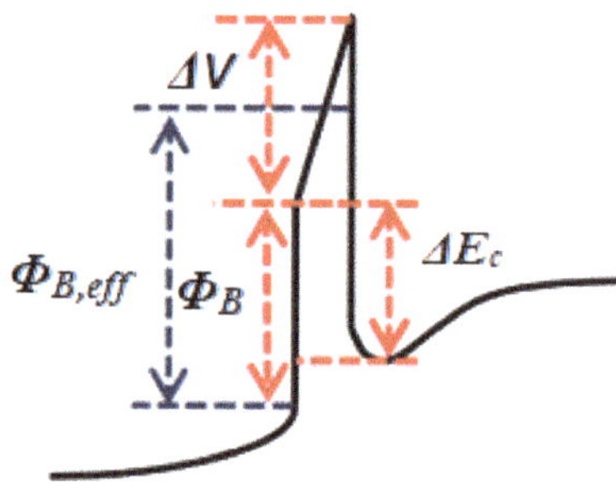

Figure 3. Construction of an effective tunnel barrier by considering the potential drop through the passivation layer.

Figure 3 can be represented by the following equation:

$$\phi_{B,eff} = \phi_B + \frac{\Delta V}{2} \quad (1)$$

where $\phi_{B,eff}$ is the effective barrier height and ΔV is the electrostatic potential drop [15]. $\phi_{B,eff}$ is 0.038 eV and 0.073 eV with a-Si:H CSFC and nc-SiOx:H CSFC layer, respectively.

Figure 4 shows the band diagram in the equilibrium condition for different CSFC layers. To observe the comparative change in the band structure, the band offsets at the valence (ΔE_v) and conduction (ΔE_c) bands of the RE-SHJ with various CSFC layers are depicted in Figure 4b,c, respectively. The values of ΔE_c are 0.082 eV and 0.046 eV for Figure 4b,c, respectively. The smaller value of ΔE_c yielded higher electron collection. It can be observed that the nc-SiO_x:H CSFC layer created a barrier for tunneling the electrons between the CSFC layer and n-type c-Si wafer. When using the nc-SiO_x:H CSFC layer, the barrier is considered triangular, as shown in Figure 4.

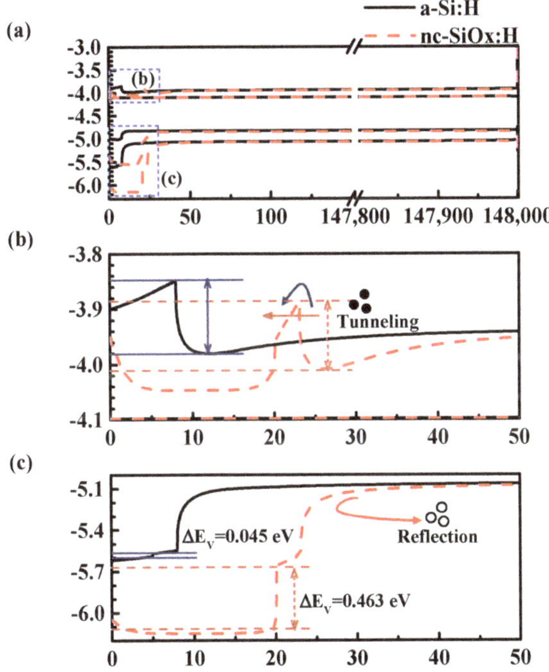

Figure 4. Band diagram of the rear emitter SHJ solar cell: (**a**) band diagram; (**b**) conduction band; (**c**) valence band with different CSFC layers.

The increase in ΔE_v acts as a hole reflector at the interface. The nc-SiO_x:H CSFC layer acts as an effective hole reflector. This leads to reduction in interface recombination. As a result, the nc-SiO_x:H CSFC layer-enhanced V_{oc} and J_{sc} are approximately 0.733 V and 40.03 mA/cm^2, respectively. To assess the effect of the nc-SiO_x:H CSFC layer, the electric field and generation/recombination rate were investigated using the numerical simulation as depicted in Figure 5. The electric field was 7.69×10^8 V/cm for the cell using the a-Si:H as passivation and the CSFC layer. For the cell using the nc-SiO_x:H CSFC layer, the electric field was increased to 2.78×10^9 V/cm between the passivation and the nc-SiO_x:H CSFC layers. The high electric field with nc-SiO_x CSFC layer is due to high band bending, as depicted in Figure 4 [16]. The nc-SiO_x:H CSFC layer can increase V_{oc} as well as lead to field effect passivation. As shown in Figure 5, the interface recombination rate increased over 1×10^{21} m^{-3}S^{-1} at the n-a-Si:H CSFC /a-Si:H passivation interface but drastically reduced under 1×10^{16} $m^{-3}S^{-1}$ between nc-SiO_x:H CSFC layer/a-Si:H passivation layer.

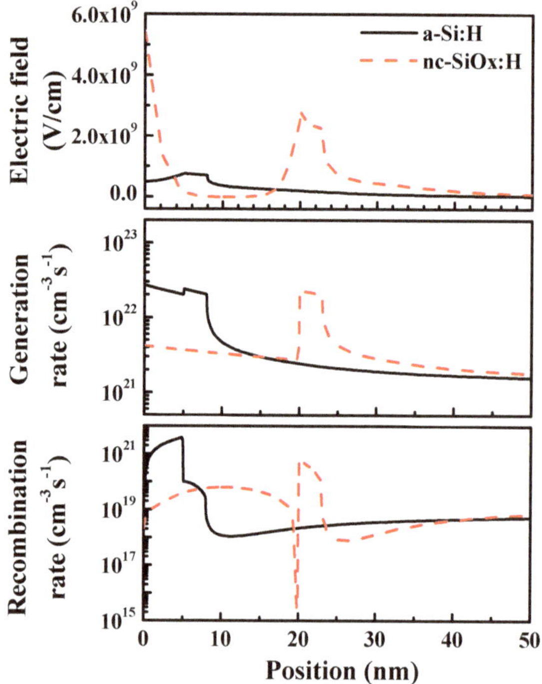

Figure 5. Influence of the CSFC layer on electric field, generation rate and recombination rate.

The field effect passivation causes decrease of recombination rate at the interface. The influence of the interface defect state of the RE-SHJ solar cell was investigated by simulation. Figure 6a shows the cell parameters as a function of the interface defect density (D_{it}) for different CSFC layers of RE-SHJ cells. When D_{it} was increased from 5×10^8 cm^{-2} to 1×10^{13} cm^{-2}, the cell efficiency and J_{sc} decreased slightly at first, and then became almost zero. These results indicate that D_{it} is required to be lower than 2×10^{10} cm^{-2} and 8×10^{10} cm^{-2} in case of using a-Si:H and nc-SiO_x:H layers, respectively, to obtain a V_{oc} over 700 mV. The V_{oc} is affected by interface recombination of the SHJ solar cell as follows [17]:

$$V_{OC} = \frac{1}{q}\left(\phi_c - AkTln\frac{qN_V S_{it}}{J_{SC}}\right) \tag{2}$$

where q is the elementary charge, ϕ_c is the effective barrier height in crystalline silicon, A is the ideality factor, k is the Boltzmann constant, T is the temperature, S_{it} is the interface recombination velocity, and N_v is the effective valence band density. The interface defect state affects surface recombination. The high D_{it} leads to increase of S_{it}, and therefore decreased V_{oc}. The interface defect state affects the quasi Fermi level and band bending. In Figure 6b, the band diagram of a SHJ solar cell using a-Si:H CSFC layer is shown with different D_{it}. When D_{it} increases, the hole quasi-Fermi energy (E_{Fp}) shifts to a high energy level, which can result in a decrease of V_{oc}. E_{Fp} can be described by the-one dimensional continuity equation as [18]:

$$E_{Fp}(x,t) = E_V(x) - kTln\frac{p(x,t)}{N_V(x)} = -q\chi(x) - q\varphi(x,t) - E_g - kTln\frac{p(x,t)}{N_V(x)} \tag{3}$$

where E_v is the valence band energy, χ is the electron affinity, $\varphi(x,t)$ is the local electric potential in the semiconductor layers and E_g is the bandgap. The value of E_{Fp} is increased due to the decrease in electric potential and an increase in N_v. The high D_{it} leads to a decrease in band bending and hence a decrease in electric potential. Additionally, it causes an improvement in the majority carrier concentration. The nc-SiO$_x$:H CSFC layer also follows the same trend as shown in Figure 6b. In contrast, the nc-SiO$_x$:H CSFC layer shows lower performance degradation even with 8×10^{10} cm^{-2} due to the tunneling effect and field effect passivation of the CSFC layer/a-Si:H passivation layer at the interface.

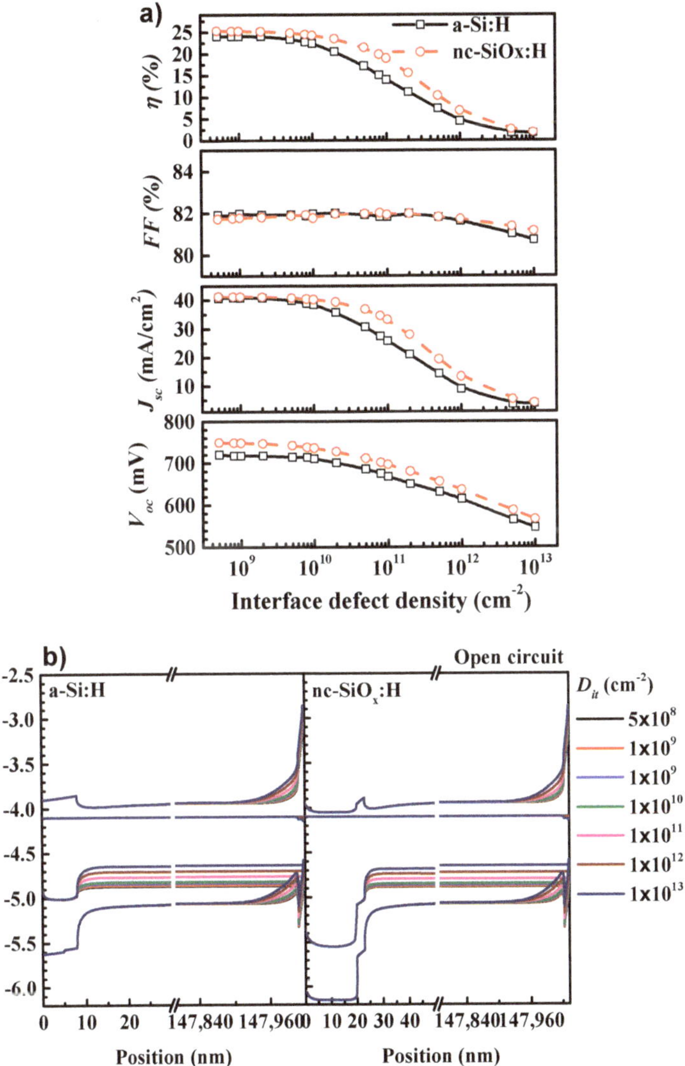

Figure 6. Schemes of rear emitter SHJ solar cells including: (**a**) The V_{oc}, J_{sc}, FF and η cell parameters as a function of interface defect density and (**b**) band diagrams with different interface defect density of a-Si:H CSFC (left) and nc-SiO$_x$:H CSFC (right) layer.

3.2. Hydrogenated Silicon Oxide Passivation Layer

The results in the previous section suggest that a low D_{it} can enhance V_{oc} and thus efficiency of the SHJ solar cell. A low D_{it} can lead to high band bending at the a-Si:H/c-Si interface, which increases the V_{oc} of the cell. A further low D_{it} can result from a hydrogenated silicon oxide (a-SiO_x:H) passivation layer formed through CO_2 plasma post-treatment. The Figure 7 shows the structural and optical properties of the a-SiO_x:H layers under CO_2 plasma post-treatment.

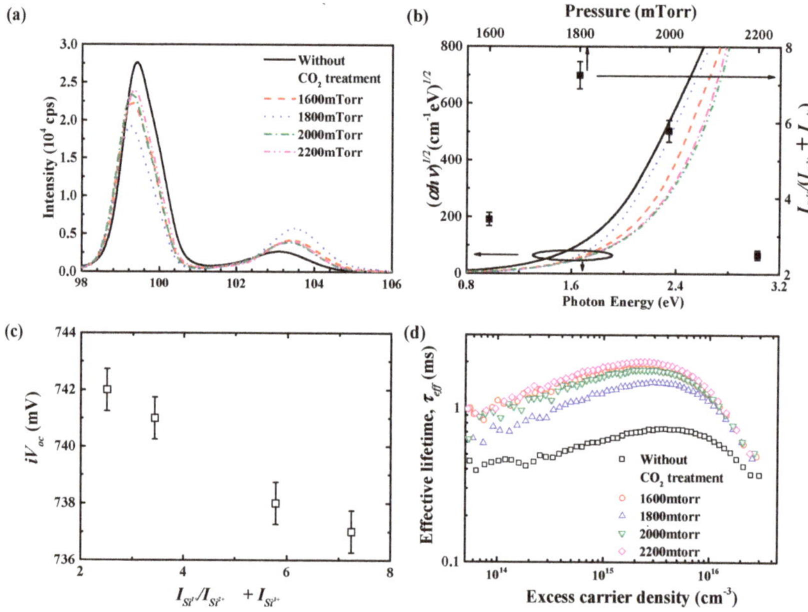

Figure 7. Structural and optical properties of a-SiO_x:H layers under different CO_2 plasma post-treatment including: (**a**) Si 2p core-level XPS spectra of a-Si:H and a-SiO_x:H different CO_2 plasma post-treatment pressure, (**b**) $I_{Si}{}^{1+}/(I_{Si}{}^{2+}+I_{Si}{}^{3+})$ ratio derived from the XPS spectra as a function of plasma post-treatment pressure and optical band gap of the a-Si:H and a-SiOx:H plotted as a function of photon energy, (**c**) iV_{oc} with respect to the $I_{Si}{}^{1+}/(I_{Si}{}^{2+}+I_{Si}{}^{3+})$ ratio, and (**d**) effective carrier lifetime measured by quasi-steady-state photo conductance (QSSPC) as a function of the treatment pressure.

Chemical composition of the a-Si:H and a-SiO_x:H layers was confirmed by employing XPS analysis, as shown in Figure 7a. After CO_2 plasma post-treatment, the pure Si peak decreased and the SiO_2 peak increased with the increase of treatment pressure. Increasing the treatment pressure led to the formation of oxygen-rich silicon oxide states. The absorption coefficient of the a-SiO_x:H layer with different CO_2 plasma treatment pressures were investigated using SE analysis, as shown in Figure 7b. The optical band gaps were determined through absorption coefficients. At higher CO_2 plasma treatment pressure, the a-SiO_x:H layer exhibited a high optical band gap, owing to the oxygen-rich components. The XPS spectrum demonstrated the increase in the formation of Si^{1+}, Si^{2+}, Si^{3+}, and Si^{4+} peaks. The XPS spectra were deconvoluted into six doublets, Si $2p_{3/2}$ (93.90 eV), Si $2P_{1/2}$ (99.90 eV), Si_2O (100.15 eV), SiO (101.05 eV), Si_2O_3 (101.80 eV) and SiO_2 (103 eV), as shown in Figure 8. In Figure 7b, the oxide state ratio ($I_{Si}{}^{1+}/(I_{Si}{}^{2+}+I_{Si}{}^{3+})$) is presented for the different working pressures during CO_2 plasma post-treatment. The oxide state ratio increased to a treatment pressure of 1800 mTorr and decreased above 1800 mTorr. The oxygen-rich components (Si^{2+} and Si^{3+}) further increased at a higher treatment pressure. To determine the optimal oxide state for high V_{oc}, as shown in Figure 7c, the effect of the oxide state ratio on iV_{oc} was investigated. The optimum value of iV_{oc} was 742 mV when the oxide state ratio was 2.5. The surface passivation properties of the a-SiO_x:H layer on textured n-type c-Si wafer was investigated. As a function of excess carrier density, the plot of τ_{eff} with a variation of plasma treatment pressure is shown in Figure 7d. After CO_2 plasma treatment, τ_{eff} increased with the increase in treatment pressure. At low injection levels of carrier density, the improvement of τ_{eff} can be explained through field effect passivation. It can be observed that a-SiO_x:H increased ΔE_v at the a-SiOx:H/c-Si interface. This phenomenon leads to enhancement of the reflection of minority carriers (hole) at the interface, which helps in low interface recombination.

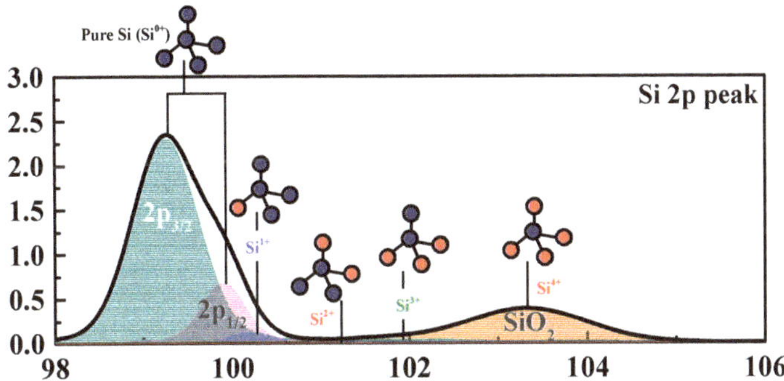

Figure 8. Peak fit for an a-SiO_x:H layer XPS spectra, showing the different oxidation states.

To confirm the cell results, we simulated the RE-SHJ solar cells using a-SiO_x:H passivation layer. In Figure 9, the J-V parameters are presented as a function of the various working pressures for the CO_2 plasma post-treatment. As depicted in Figure 9, the efficiency increases with CO_2 plasma post-treatment. The optical band gap of the a-SiO_x:H layer increased with CO_2 plasma treatment. This optical band gap created a higher ΔE_v at the a-SiO_x:H/c-Si interface, which may have assisted in hole reflection and decreased recombination. We investigated the effect of using a-SiO_x:H passivation layer. This is because the a-SiO_x:H has a wider band gap and good passivation properties. The result shows that such an increase of the V_{oc} is feasible by employing an appropriate a-SiO_x:H as the passivation layer.

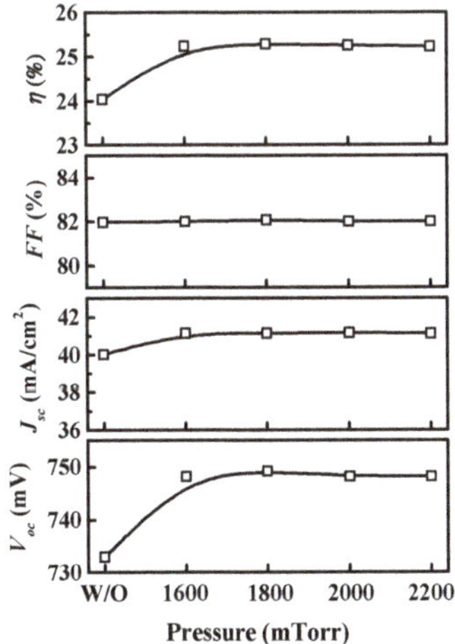

Figure 9. Estimated J-V parameters of rear emitter HIT cells as a function of CO_2 plasma post-treatment.

4. Conclusions

In this paper, the impact of various CSFC layers and interface defect densities on the efficiency of RE-SHJ solar cells was discussed through simulations and experiments. To compare the effect of the various CSFC layers, simulations were performed using a-Si:H and nc-SiO_x:H CSFC materials. Using a nc-SiO_x:H CSFC layer results in the decreased parasitic absorption and recombination loss through field effect passivation, thus enhancing RE-SHJ cell efficiency. The output parameters of RE-SHJ solar cell, such as V_{oc} = 733 mV, J_{sc} = 40.03 mA/cm^2, FF = 81.97% and η = 24.04%, were obtained using a nc-SiO_x:H CSFC layer. The increase in interface defect states resulted in a decrease in band bending and a shift in E_{Fp} to high energy and hence a reduction in V_{oc}. Finally, the nc-SiO_x:H CSFC layer has better performance in the same interface defect state than the a-Si:H CSFC layer. This improvement is attributed to the tunneling effect and field effect passivation when using the nc-SiO_x:H CSFC layer. To improve the surface passivation quality of the RE-SHJ cell, the a-SiO_x:H layer was used as the passivation layer. The V_{oc} increased due to the decrease of the recombination rate at the interface. This investigation suggests that a suitable a-SiO_x:H passivation layer can significantly improve the V_{oc} and efficiency of SHJ solar cells.

Author Contributions: S.L. is the main author. S.L. and D.P.P. carried out the investigations and undertook the simulation and data analysis. Y.K., E.-C.C. and J.Y. supervised the project as was also responsible for funding acquisition. S.L. and J.P. were responsible for the writing—original draft preparation while all authors were involved in writing—review and editing. All authors have read and agreed to the published version of the manuscript.

Funding: This work was supported by the Korea Institute of Energy Technology Evaluation and Planning (KETEP) and the Ministry of Trade, Industry and Energy (MOTIE) of the Republic of Korea (No. 20203030010310) and the National Research Foundation of Korea(NRF) funded by the Korea government (MSIT) (NRF-2019R1A2C1009126).

Conflicts of Interest: The authors declare no conflict of interest.

References

1. Zheng, C.; Kammen, D.M. An innovation-focused roadmap for a sustainable global photovoltaic industry. *Energy Policy* **2014**, *67*, 159–169. [CrossRef]
2. Battaglia, C.; Cuevas, A.; De Wolf, S. High-efficiency crystalline silicon solar cells: Status and perspectives. *Energy Environ. Sci.* **2016**, *9*, 1552–1576. [CrossRef]
3. Yoshikawa, K.; Kawasaki, H.; Yoshida, W.; Irie, T.; Konishi, K.; Nakano, K.; Uto, T.; Adachi, D.; Kanematsu, M.; Uzu, H.; et al. Silicon heterojunction solar cell with interdigitated back contacts for a photoconversion efficiency over 26%. *Nat. Energy* **2017**, *2*, 17032. [CrossRef]
4. Tohoda, S.; Fujishima, D.; Yano, A.; Ogane, A.; Matsuyama, K.; Nakamura, Y.; Tokuoka, N.; Kanno, H.; Kinoshita, T.; Sakata, H.; et al. Future directions for higher-efficiency HIT solar cells using a Thin Silicon Wafer. *J. Non-Cryst. Solids* **2012**, *358*, 2219–2222. [CrossRef]
5. Deligiannis, D.; Van Vliet, J.; Vasudevan, R.; Van Swaaij, R.A.C.M.M.; Zeman, M. Passivation mechanism in silicon heterojunction solar cells with intrinsic hydrogenated amorphous silicon oxide layers. *J. Appl. Phys.* **2017**, *121*, 085306. [CrossRef]
6. Liu, W.; Zhang, L.; Cong, S.; Chen, R.; Wu, Z.; Meng, F.; Shi, Q.; Liu, Z. Controllable a-Si:H/c-Si interface passivation by residual SiH4 molecules in H2 plasma. *Sol. Energy Mater. Sol. Cells* **2018**, *174*, 233–239. [CrossRef]
7. Holman, Z.C.; Descoeudres, A.; Barraud, L.; Fernandez, F.Z.; Seif, J.P.; De Wolf, S.; Ballif, C. Current losses at the front of silicon heterojunction solar cells. *IEEE J. Photovolt.* **2012**, *2*, 7–15. [CrossRef]
8. Mazzarella, L.; Kirner, S.; Stannowski, B.; Korte, L.; Rech, B.; Schlatmann, R. P-type microcrystalline silicon oxide emitter for silicon heterojunction solar cells allowing current densities above 40 mA/cm2. *Appl. Phys. Lett.* **2015**, *106*, 023902. [CrossRef]
9. Taguchi, M.; Terakawa, A.; Maruyama, E.; Tanaka, M. Obtaining a higher voc in HIT cells. *Prog. Photovolt. Res. Appl.* **2005**, *13*, 481–488. [CrossRef]
10. Bivour, M.; Schröer, S.; Hermle, M. Numerical analysis of electrical TCO/a-Si:H(p) contact properties for silicon heterojunction solar cells. *Energy Procedia* **2013**, *38*, 658–669. [CrossRef]

11. Varache, R.; Kleider, J.P.; Gueunier-Farret, M.E.; Korte, L. Silicon heterojunction solar cells: Optimization of emitter and contact properties from analytical calculation and numerical simulation. *Mater. Sci. Eng. B Solid-State Mater. Adv. Technol.* **2013**, *178*, 593–598. [CrossRef]
12. Bivour, M.; Schröer, S.; Hermle, M.; Glunz, S.W. Silicon heterojunction rear emitter solar cells: Less restrictions on the optoelectrical properties of front side TCOs. *Sol. Energy Mater. Sol. Cells* **2014**, *122*, 120–129. [CrossRef]
13. Mazzarella, L.; Morales-Vilches, A.B.; Korte, L.; Schlatmann, R.; Stannowski, B. Ultra-thin nanocrystalline n-type silicon oxide front contact layers for rear-emitter silicon heterojunction solar cells. *Sol. Energy Mater. Sol. Cells* **2018**, *179*, 386–391. [CrossRef]
14. Richter, A.; Smirnov, V.; Lambertz, A.; Nomoto, K.; Welter, K.; Ding, K. Versatility of doped nanocrystalline silicon oxide for applications in silicon thin-film and heterojunction solar cells. *Sol. Energy Mater. Sol. Cells* **2018**, *174*, 196–201. [CrossRef]
15. Varache, R.; Leendertz, C.; Gueunier-Farret, M.E.; Haschke, J.; Muñoz, D.; Korte, L. Investigation of selective junctions using a newly developed tunnel current model for solar cell applications. *Sol. Energy Mater. Sol. Cells* **2015**, *141*, 14–23. [CrossRef]
16. Gudovskikh, A.S.; Ibrahim, S.; Kleider, J.P.; Damon-Lacoste, J.; Roca i Cabarrocas, P.; Veschetti, Y.; Ribeyron, P.J. Determination of band offsets in a-Si:H/c-Si heterojunctions from capacitance-voltage measurements: Capabilities and limits. *Thin Solid Films* **2007**, *515*, 7481–7485. [CrossRef]
17. Dao, V.A.; Heo, J.; Choi, H.; Kim, Y.; Park, S.; Jung, S.; Lakshminarayan, N.; Yi, J. Simulation and study of the influence of the buffer intrinsic layer, back-surface field, densities of interface defects, resistivity of p-type silicon substrate and transparent conductive oxide on heterojunction with intrinsic thin-layer (HIT) solar cell. *Sol. Energy* **2010**, *84*, 777–783. [CrossRef]
18. Bashiri, H.; Karami, M.A.; Mohammadnejad, S. A theoretical investigation of quantum confinement effects in heterojunction silicon solar cells. *Indian J. Phys.* **2018**, *92*, 349–356. [CrossRef]

Article

An Analysis of Fill Factor Loss Depending on the Temperature for the Industrial Silicon Solar Cells

Kwan Hong Min [1,2], Taejun Kim [3], Min Gu Kang [1], Hee-eun Song [1], Yoonmook Kang [4], Hae-Seok Lee [4], Donghwan Kim [2], Sungeun Park [1,*] and Sang Hee Lee [1,*]

1 Photovoltaics Laboratory, Korea Institute of Energy Research, Daejeon 34129, Korea; steel1217@nate.com (K.H.M.); mgkang@kier.re.kr (M.G.K.); hsong@kier.re.kr (H.-e.S.)
2 Department of Materials Science and Engineering, Korea University, Seoul 02841, Korea; solar@korea.ac.kr
3 Hyundai Energy Solutions, PV R&D Center, Eumseong, Cheongju 27711, Korea; taejunkim@hyundai-es.co.kr
4 Department of Energy Environment Policy and Technology, Green School, Graduate School of Korea Energy and Environment, Korea University, Seoul 02841, Korea; ddang@korea.ac.kr (Y.K.); lhseok@korea.ac.kr (H.-S.L.)
* Correspondence: separk@kier.re.kr (S.P.); aktkd88@naver.com (S.H.L.); Tel.: +82-42-860-3419 (S.P.); +82-42-860-3419 (S.H.L.)

Received: 12 May 2020; Accepted: 3 June 2020; Published: 7 June 2020

Abstract: Since the temperature of a photovoltaic (PV) module is not consistent as it was estimated at a standard test condition, the thermal stability of the solar cell parameters determines the temperature dependence of the PV module. Fill factor loss analysis of crystalline silicon solar cell is one of the most efficient methods to diagnose the dominant problem, accurately. In this study, the fill factor analysis method and the double-diode model of a solar cell was applied to analyze the effect of J_{01}, J_{02}, R_s, and R_{sh} on the fill factor in details. The temperature dependence of the parameters was compared through the passivated emitter rear cell (PERC) of the industrial scale solar cells. As a result of analysis, PERC cells showed different temperature dependence for the fill factor loss of the J_{01} and J_{02} as temperatures rose. In addition, we confirmed that fill factor loss from the J_{01} and J_{02} at elevated temperature depends on the initial state of the solar cells. The verification of the fill factor loss analysis was conducted by comparing to the fitting results of the injection dependent-carrier lifetime.

Keywords: fill factor loss analysis; double-diode model; PERC; temperature dependence; recombination current density; parasitic resistance

1. Introduction

The fill factor (*FF*), open circuit voltage (V_{oc}) and short circuit current (J_{sc}) of a solar cell are important parameters because they determine the maximum power that a solar cell can generate. It is well known that the fill factor of silicon solar cells is influenced by the recombination current and parasitic resistance. In order to clarify the effect of these factors on a fill factor, the double-diode model is generally used as a theoretical concept. In the double-diode model, the photo-generated current density reduces due to the saturation current densities (J_{01}, J_{02}), and parasitic resistance (R_s, R_{sh}) as the equation shows below.

$$J = J_{ph} - J_{01}\left[\exp\left\{\frac{q(V+JR_s)}{n_1kT}\right\} - 1\right] - J_{02}\left[\exp\left\{\frac{q(V+JR_s)}{n_2kT}\right\} - 1\right] - \frac{(V+JR_s)}{R_{sh}}, \quad (1)$$

where J is the diode circuit current density, J_{ph} is the photo-generated current density, J_{01} is the recombination current density in quasi-neutral region, V is the diode voltage, q is the elementary charge, k is Boltzmann's constant, n_1 is the ideality factor for the diode 1, n_2 is the ideality factor for the diode 2, and T is the absolute temperature. The J_{01} in the equation is the information of the recombination current from the base material (quasi-neutral region) and both side surface of the solar cell. J_{02} generally describes the Shockley-Read-Hall (SRH) recombination at the p-n junction (space charge region), but there are also other recombination sources such as edge recombination [1] and localized high defect region [2,3].

Analyzing the recombination and resistive loss of a solar cell from the perspective of fill factor is a very efficient method because it directly shows the gain of the conversion efficiency from the loss factor. Hence, there were some effort to extract fill factor loss at the space charge region (J_{02}) by determining fill factor loss from J_{01} (FF_{J01}) and subtracting the pseudo fill factor (pFF) that is not influenced by the series resistance [4,5]. Afterward, A. Khanna proposed a method to quantify the fill factor loss of silicon solar cells due to parasitic resistance as well as J_{02}, because there was a limitation that the shunt resistance must be extremely high to extract an exact fill factor loss due to J_{02} in this method. In addition, since this approach does not need to assume n_2 value during the fitting procedure, it showed that a more accurate analysis is possible. The theoretical derivation is well described in [6].

This fill factor loss analysis was used in this experiment to investigate the source of fill factor loss depending on the elevated temperature, because the module efficiency assessed at the standard test condition (STC, 25 °C) reduces at the actual operating temperature [7]. Figure 1 shows the measured PV module temperature change according to the seasonal temperature in Daejeon and Seoul of South Korea. The PV module temperature rises to 60 °C which will highly degrade the module power when it is operating. The V_{oc} reduction is well known for the main source of efficiency loss. V_{oc} is a function of the saturation current (I_0) which is strongly dependent on the temperature as it can be seen in Equations (2) and (3) [8]. Where q is the elementary charge, A is the area, D is the diffusion constant, n_i is the intrinsic carrier concentration, L is the diffusion length, N_D is the doping concentration, and T is kelvin temperature. By extracting the fill factor loss of each parameter (J_{01}, J_{02}, R_s, and R_{sh}), it will be able to provide an accurate diagnosis and appropriate solution to reduce the temperature dependence of the PV module.

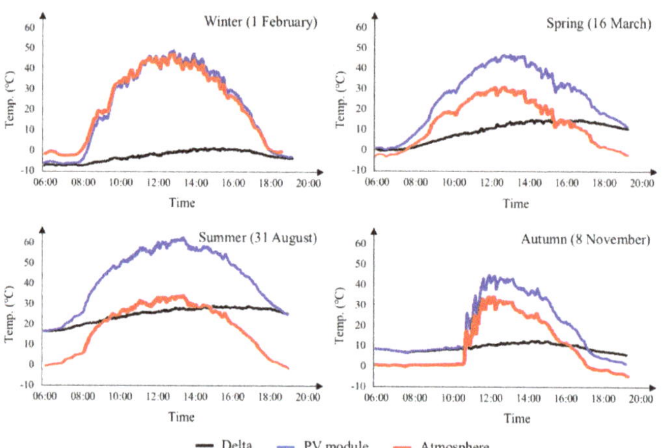

Figure 1. Photovoltaic (PV) module temperature change according to seasonal temperature.

For the analysis of fill factor loss, industrial passivated emitter and rear cell (PERC), which is leading product in PV industries, was used [9]. Many researches for optimization of both side surface passivation [10–16] and laser ablation on rear aluminum oxide (AlO_x) layer [17,18] for the local back surface field formation (LBSF) [19] have improved efficiency of the PERC and reached 22.8% [20]. In this paper, temperature dependence of two industrial type PERC samples were compared to inspect dominant factor of fill factor loss.

$$I_0 = qA\frac{Dn_i^2}{LN_D}, \tag{2}$$

$$n_i(T) = 9.15 \times 10^9 \left(\frac{T + 273.15}{300}\right)^2 \exp\left(\frac{-6880}{T + 273.15}\right). \tag{3}$$

2. Experimental

2.1. Analysis Process of Fill Factor Loss

Figure 2 briefly summarizes the steps of the fill factor loss analysis. The analysis of the fill factor loss starts with the calculation of FF_{J01}. The FF_{J01} can be determined when the effect of J_{02}, R_s and R_{sh} is absent. With this assumption, Equation (1) changes to Equation (4).

$$J = J_{ph} - J_{01}\left[\exp\left\{\frac{qV}{nkT}\right\} - 1\right]. \tag{4}$$

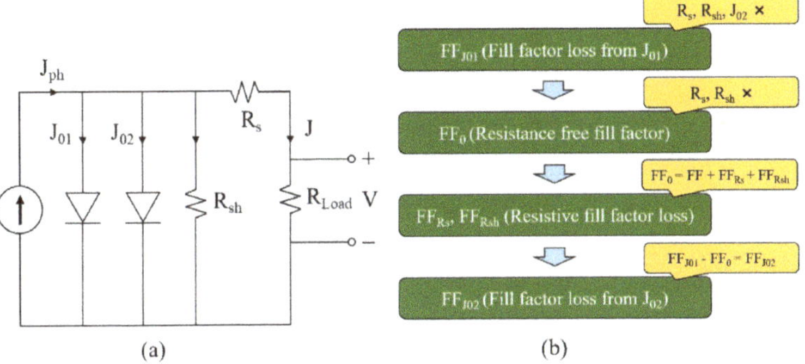

Figure 2. (**a**) Equivalent circuit of solar cells with double-diode model, and (**b**) process of fill factor loss analysis.

By imposing the boundary condition $J = 0$ at $V = V_{oc}$ and $J = J_{sc}$ at $V = 0$, relationship between J and V can be expressed in Equation (5), which FF_{J01} is allowed to be obtained.

$$J = J_{sc} - \frac{J_{sc}}{\exp\left(\frac{qV_{oc}}{nkT}\right) - 1}\left[\exp\left\{\frac{qV}{nkT}\right\} - 1\right]. \tag{5}$$

Afterwards, resistance free fill factor (FF_0), which does not have loss from the parasitic resistance, is derived as in Equation (6) by considering the double-diode model at maximum power point (MPP). Where, J_{mpp} and V_{mpp} is the current density and voltage at power maximum point, respectively. Equation (6) is composed of three terms: real fill factor, fill factor loss from R_s (FF_{Rs}), and fill factor loss from R_{sh} (FF_{Rsh}), respectively from the left term.

$$FF_0 = FF + \frac{J_{mpp}^2 R_s}{V_{oc} J_{sc}} + \frac{\left(V_{mpp} + J_{mpp} R_s\right)^2}{R_{sh} V_{oc} J_{sc}}. \tag{6}$$

As a last step, the fill factor loss from J_{02} can be readily determined from the difference of FF_{J01} and FF_0 Equation (7), because the difference comes from the absence of the J_{02} effect.

$$FF_{J02} = FF_{J01} - FF_0. \tag{7}$$

2.2. Sample Measurements with Elevated Temperature

Two industrial solar cells with PERC structure (Figure 3) were prepared as samples (PERC A and B). The PERC A and B were fabricated on different production lines, but the front contact pattern was unified to exclude optical loss due to the front contact. In case of the PERC A, silver paste for the front contact contains higher glass frit contents compared to the PERC B. The characteristics of illuminated *J-V* were measured by using an in-house solar simulator, which can control the temperature of substrate. Even though this fill factor analysis method assumed that J_{01} follows an n factor of 1, there were changes in the n factor depending on the temperature. Therefore, n factor at 1-sun, extracted from the Suns-V_{oc} measurement, was used for analysis. The temperature of cells during the Suns-V_{oc} measurement was controlled by using a box that has an ambient heater. The conditions of the experimental temperature were set to 25 °C, 45 °C, and 65 °C.

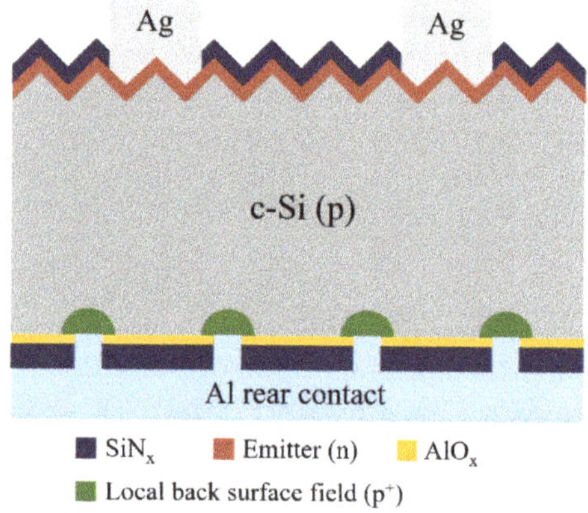

Figure 3. Schematic of the industrial PERC solar cell structure.

3. Results and Discussion

3.1. Temperature Dependence

The two PERC solar cell samples were named by the PERC A and B. As a result of the illuminated *J-V* measurements (Table 1), V_{oc} and fill factor of PERC B were 17 mV and 1.93 % higher at 25 °C, respectively. J_{sc} difference was not significant (~1.3 mA/cm^2) and it comes from the different fraction of front metal contact. Graphs in Figure 4 show the drop of V_{oc} and fill factor (depending on temperature), and also presents the average temperature coefficient as well. The average drop of V_{oc} and fill factor with the PERC A cell were slightly higher than the PERC B cell. Especially, the larger fill factor drops occurred at 45 °C for the PERC A and 65 °C for the PERC B. Illuminated *J-V* measurements alone do not have enough evidence to accurately diagnose what the main problem is.

Table 1. Results of illuminated *J-V* measurement and n factor (from Suns-V_{oc}) of PERC A and B according to temperature dependence.

Parameters	PERC A			PERC B		
Temperature [°C]	25	45	65	25	45	65
V_{oc} [mV]	654	616	578	671	635	596
J_{sc} [mA/cm^2]	38.49	38.74	39.05	39.79	40.08	40.21
V_{mpp} [mV]	564	517	478	580	526	494
J_{mpp} [mA/cm^2]	35.84	35.92	36.18	37.85	38.07	38.20
FF [%]	80.28	77.70	76.60	82.21	80.53	78.62
R_s [Ω·cm^2]	1.58	1.77	1.631	1.49	1.42	1.60
R_{sh} [Ω·cm^2]	10252	2796	2982	9553	26562	5126
η [%]	20.21	18.57	17.28	21.96	20.49	18.87
n factor (Suns-V_{oc})	1.07	1.14	1.17	1.05	1.12	1.19

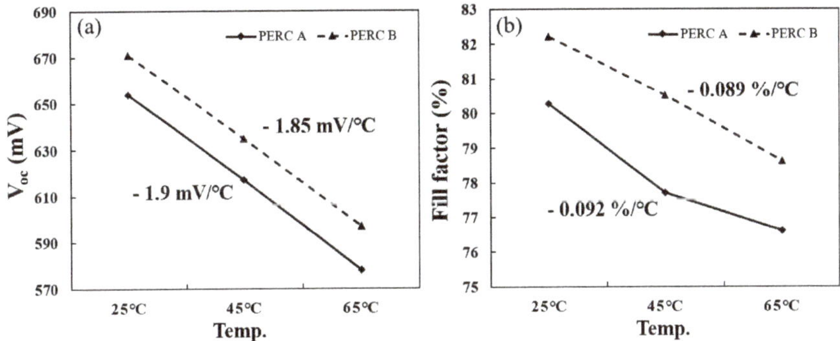

Figure 4. Temperature coefficient of (**a**) V_{oc} and (**b**) fill factor.

3.2. Results of Fill Factor Loss Analysis

Table 2 organizes the values of each fill factor loss according to the parameters from double-diode model. From this data, Figures 5 and 6 shows the effect of each parameter on the total fill factor loss. At initial temperature, percentage of FF_{J01} was higher than FF_{J02} for the PERC B cell. On the other hand, FF_{J02} had a higher percentage of total fill factor loss than FF_{J01} for the PERC A cell. As the temperature rose, FF_{J01} ratio of both PERC A and B increased while FF_{J02} ratio decreased. This can be explained by the temperature dependence of J_{01} and J_{02} in Equations (8) and (9).

$$J_{01}(T) = J_{01}(25\ °\mathrm{C}) \times \left(\frac{n_i(T)}{n_i(25\ °\mathrm{C})}\right)^2, \tag{8}$$

$$J_{02}(T) = J_{02}(25\ °\mathrm{C}) \times \left(\frac{n_i(T)}{n_i(25\ °\mathrm{C})}\right). \tag{9}$$

J_{01} is proportional to square of $n_i(T)$, which increases sensitively according to the temperature, while J_{02} is directly proportional to $n_i(T)$. Due to the difference of the temperature dependence, the effect of J_{01} becomes dominate as temperature goes up. However, the FF_{J02} ratio and the absolute value are still higher than FF_{J01} at 65 °C for the PERC A cell. A possible explanation is that the initial value of FF_{J02} (1.86%) was twice higher than that of FF_{J01} (0.82%) and helped increase at a high temperature.

Table 2. Results of fill factor loss analysis of PERC A and PERC B according to temperature (the values are absolute).

FF Loss Analysis	PERC A			PERC B		
Temperature [°C]	25	45	65	25	45	65
Ideal *FF* [%]	83.78	82.89	82.36	84.08	83.43	82.67
Total *FF* loss [%]	3.50	5.19	5.78	1.88	2.91	4.06
FF_{J01} [%]	0.82	1.70	2.10	0.58	1.42	2.30
FF_{J02} [%]	1.86	2.47	2.69	0.48	0.70	0.75
FF_{Rs} [%]	0.81	0.97	0.94	0.80	0.78	0.98
FF_{Rsh} [%]	0.01	0.05	0.05	0.02	0.01	0.03
FF_0 [%]	81.00	78.72	77.57	83.02	81.31	79.62
Real *FF* (LIV) [%]	80.28	77.70	76.60	82.21	80.53	78.62

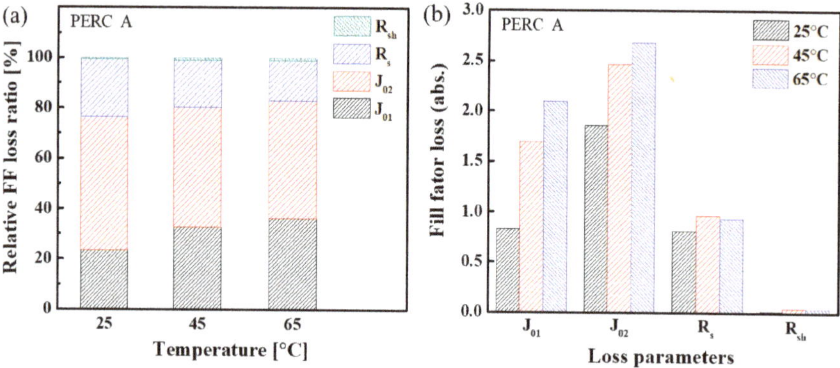

Figure 5. Fill factor loss analysis results of PERC A: (**a**) Normalized fill factor ratio; (**b**) Absolute fill factor loss.

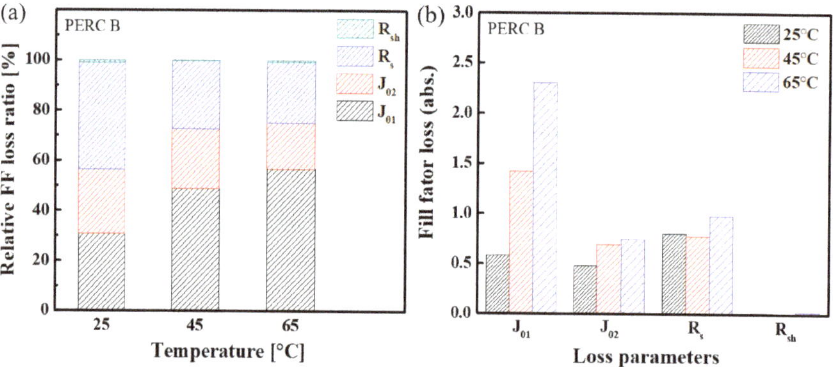

Figure 6. Fill factor loss analysis results of PERC B: (**a**) Normalized fill factor ratio; (**b**) Absolute fill factor loss.

In addition, the temperature dependence of FF_{J02} for the PERC B cell (Figure 6b) reveals that the FF_{J02} will not increase significantly at elevated temperature if the initial ratio to the total fill factor loss is relatively small. On the other hand, the difference of FF_{J01} between 25 °C and 65 °C increased up to 20%, even though the initial difference was only 7%. This means that the effect of J_{01} on fill factor will be accelerated if the initial loss is relatively high.

In case of the PERC B cell, one can diagnose that the recombination of emitter, quasi-neutral region, or surface passivation have to be improved. Reduction of FF_{Rs} ratio for the PERC B cell will further reduce the temperature dependence, even though there were small variation at the elevated temperature. The ratio and temperature dependence of FF_{Rsh} were negligible for both PERC A and B cells.

3.3. *Verification Through Injection Dependent-Carrier Lifetime*

In order to verify the results of the fill factor loss analysis, the lifetime was measured through Sinton instrument's Suns-V_{oc} equipment at 25 °C. The recombination current density of the samples was evaluated by using a lifetime fitting method. Through the specific resistance and thickness of the wafer, the Auger recombination and the radiative recombination are represented as lifetime according to the excess carrier concentration as graphs in Figure 7 show.

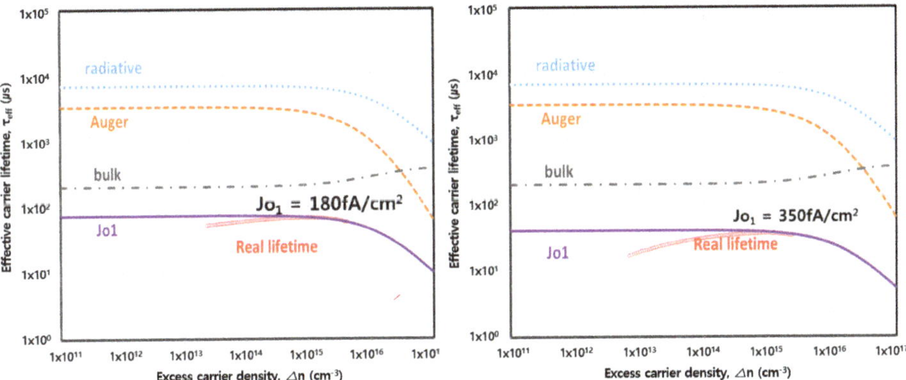

Figure 7. Fitting results of injection dependent-carrier lifetime for the PERC A (**Left**) and B (**Right**).

The expressions for the intrinsic carrier lifetime and bulk lifetime can be expressed as follows.

$$\tau = \frac{\Delta n}{R} \quad (10)$$

$$\frac{1}{\tau_{intrinsic+bulk}} = \frac{1}{\tau_{rad}} + \frac{1}{\tau_{Auger}} + \frac{1}{\tau_{SRH}} \quad (11)$$

Here, intrinsic recombination can be expressed as an Equation (12) [21].

$$R_{Intrinsic} = R_{Auger} + R_{Radiative} = np\left(1.8\times10^{-24}{n_0}^{0.65} + 6\times10^{-25}{p_0}^{0.65} + 3\times10^{-27}\Delta n^{0.8} + 9.5\times10^{-15}\right) \quad (12)$$

Where n and p are the free electron and hole concentrations, n_0 and p_0 are the electron and hole concentration in thermal equilibrium, and Δn is the excess carrier densities. The bulk lifetime of a wafer can be measured with quasi-steady-state photoconductance (QSSPC) equipment. To measure the bulk lifetime of the wafer, the doped layer was removed from the wafer using KOH solution after $POCl_3$ diffusion, and then the Al_2O_3/SiN_x stack passivation layer was deposited on both sides. Bulk lifetime can be calculated by the following two formulas [22,23].

$$J_o = \frac{S_{eff}}{qn_i^2} \times (N_A + \Delta n), \quad (13)$$

$$\frac{1}{\tau_{bulk}} = \frac{1}{\tau_{eff}} - \frac{2S_{eff}}{W}, \quad (14)$$

where W is the sample thickness, S_{eff} is the surface recombination velocity. The S_{eff} values were calculated as 9.8 cm/s at the 10^{15} cm^{-3} injection level which is commonly used [22]. As a result of our calculation, the bulk lifetime was 343 µs for our wafer. The graphs in Figure 7 shows these three lifetimes of radiative, Auger and bulk recombination. If the parameter of J_{01} is added to this formula, lifetime graph at high injection can be fit by using the formula below:

$$\frac{1}{\tau_{eff}} - \frac{1}{\tau_{intrinsic}} = \frac{1}{\tau_{bulk}} + \frac{J_{01}(N_A + \Delta n)}{q{n_i}^2 w}. \quad (15)$$

By fitting the lifetime graph through the J_{01} value, the J_{01} value was calculated as 350 fA/cm^2 for the PERC A cell and 180 fA/cm^2 for the PERC B cell. Similar to the results of the fill factor loss analysis, the J_{01} value was significantly higher for the PERC A cell.

Furthermore, we could draw the theoretical graph of FF_0 as function of J_{02} (25 °C) by applying the J_{01} value from Figure 7 to the J-V equation of double diode model without resistance Equation (16). The J_{02} (25 °C) of the PERC A cell was extracted by finding the point that the FF_0 from the fill factor loss analysis equal to the graph (black line in Figure 8). In this way, the J_{02} (25 °C) of the PERC A and B cells were assessed as 27 nA/cm^2 and 7.35 nA/cm^2, respectively. The difference of the J_{02} value was also expected from the graph in Figure 7. The graphs generally showed poor fitting results at low-level injection region. This means that the PERC A cell has poor properties of J_{02} or shunt resistance because these are well known as main reasons for low lifetime of solar cells in the low-level injection region [24].

$$J = J_{ph} - J_{01}\left[\exp\left\{\frac{qV}{n_1kT}\right\} - 1\right] - J_{02}\left[\exp\left\{\frac{qV}{n_2kT}\right\} - 1\right]. \tag{16}$$

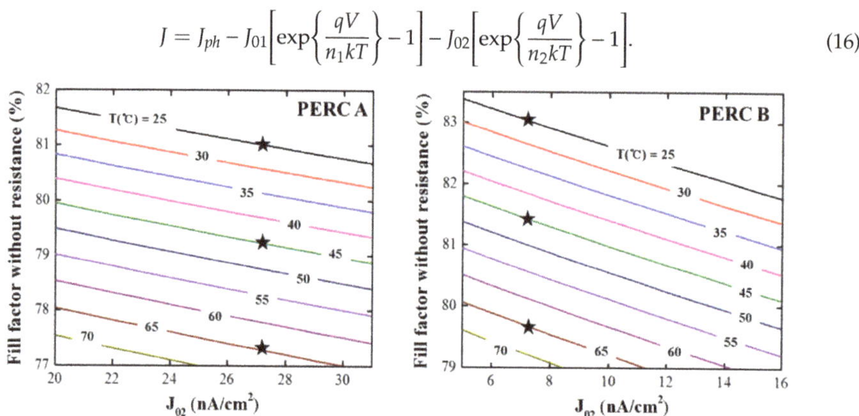

Figure 8. Theoretical calculation of FF_0 as function of the J_{02} (25 °C). The star marks are the FF_0 value extracted from fill factor loss analysis (Table 2).

In case of the temperature above 25 °C, J_{01} and J_{02} values were calculated by using Equations (8) and (9). Through these J_{01} and J_{02} values, temperature conditions of the J_{02} (25 °C)-FF_0 graph were extended from 30 to 70 °C. We confirmed that the FF_0 from Table 2 matched quite well with the theoretical graph. Consequently, we confirmed that the fill factor loss analysis was conducted precisely by checking the theoretically extracted J_{01} and J_{02} values.

4. Conclusions

We conducted fill factor loss analysis to investigate the temperature dependence of industrial PERC cells based on the double diode model. While analyzing the loss of fill factor for the industrial PERC A and B cells, it was confirmed that both samples showed temperature dependence of fill factor (~0.092%/ °C). However, the main parameter that dominantly contribute to temperature dependence were different depending on the initial state of the samples.

For PERC A, J_{02} was the main parameter of fill factor loss in initial state (25 °C). However, the contribution of the J_{01} to the fill factor loss increased as the temperature rose. It could be estimated that the PERC A cell has potential to improve fill factor by optimizing J_{02} or shunt resistance. On the other hand, PERC B has a higher FF_{Rs} ratio at initial state (25 °C) than other parameters, but the ratio of FF_{Rs} reduced due to the rapid increase in the FF_{J01}. Therefore, we believe that the J_{01} has to be reduced for the PERC B cell by optimizing the doping process, passivation quality, and metal contacts.

In addition, we verified the accuracy of the fill factor loss analysis by comparing it to the fitting results of the injection dependent-carrier lifetime. The J_{01} and J_{02} values at 25 °C were extracted, and it was confirmed that the values of FF_0 obtained from the fill factor loss analysis were in good agreement with the theoretically calculated values. In conclusion, it seems very important to reduce initial fill factor loss from the J_{01}, because it has a high temperature dependence. In case of the J_{02} and R_s, they seem less detrimental for the reduction of fill factor at an elevated temperature. Also, we confirmed that the fill factor loss analysis used in this experiment is very efficient to diagnose the temperature dependence of the solar cells, as well as their initial state.

Author Contributions: K.H.M. designed experiments, performed the experiments and wrote manuscripts; Formal analysis, T.K.; data curation, M.G.K.; investigation, H.-e.S.; review and editing, Y.K. and H.-S.L.; supervision, D.K.; project administration and supervision, S.P.; writing-review and editing and project administration, S.H.L. All authors have read and agreed to the published version of the manuscript.

Funding: This research was funded by the Ministry of Knowledge Economy, Korea (Project No. 20183030019460), and funded by the Ministry of Science, ICT & Future Planning (Project No. 2017M1A2A2086911).

Acknowledgments: This work was conducted under the New and Renewable Energy Technology Development Program of the Korea Institute of Energy Technology Evaluation and Planning (KETEP) through a grant funded by the Ministry of Knowledge Economy, Korea (Project No. 20183030019460), and by the Technology Development Program to Solve Climate Changes of the National Research Foundation (NRF) funded by the Ministry of Science, ICT & Future Planning (Project No. 2017M1A2A2086911).

Conflicts of Interest: The authors declare no conflict of interest. Also, the funders had no role in the design of the study; in the collection, analyses, or interpretation of data; in the writing of the manuscript, or in the decision to publish the results.

References

1. McIntosh, K. Lumps, humps and bumps: Three detrimental effects in the current–voltage curve of silicon solar cells, in: Centre for Photovoltaic Engineering. *Univ. N. S. W. Aust.* **2001**, 171.
2. Breitenstein, O.; Rakotoniaina, J.; Al Rifai, M.H.; Werner, M. Shunt types in crystalline silicon solar cells. *Prog. Photovolt. Res. Appl.* **2004**, *12*, 529–538. [CrossRef]
3. Steingrube, S.; Breitenstein, O.; Ramspeck, K.; Glunz, S.; Schenk, A.; Altermatt, P.P. Explanation of commonly observed shunt currents in c-Si solar cells by means of recombination statistics beyond the Shockley-Read-Hall approximation. *J. Appl. Phys.* **2011**, *110*, 014515. [CrossRef]
4. Greulich, J.; Glatthaar, M.; Rein, S. Fill factor analysis of solar cells' current–voltage curves. *Prog. Photovolt. Res. Appl.* **2010**, *18*, 511–515. [CrossRef]
5. Hoenig, R.; Glatthaar, M.; Clement, F.; Greulich, J.; Wilde, J.; Biro, D. New measurement method for the investigation of space charge region recombination losses induced by the metallization of silicon solar cells. *Energy Procedia* **2011**, *8*, 694–699. [CrossRef]
6. Khanna, A.; Mueller, T.; Stangl, R.A.; Hoex, B.; Basu, P.K.; Aberle, A.G. A fill factor loss analysis method for silicon wafer solar cells. *IEEE J. Photovolt.* **2013**, *3*, 1170–1177. [CrossRef]
7. Correa-Betanzo, C.; Calleja, H.; De León-Aldaco, S. Module temperature models assessment of photovoltaic seasonal energy yield. *Sustain. Energy Technol. Assess.* **2018**, *27*, 9–16. [CrossRef]
8. Sproul, A.; Green, M. Improved value for the silicon intrinsic carrier concentration from 275 to 375 K. *J. Appl. Phys.* **1991**, *70*, 846–854. [CrossRef]
9. The International Technology Roadmap for Photovoltaic (ITRPV). Available online: https://itrpv.vdma.org (accessed on 8 May 2020).
10. Glunz, S.W. High-efficiency crystalline silicon solar cells. *Adv. Optoelectron.* **2007**, *2007*, 1–15. [CrossRef] [PubMed]
11. Cuevas, A.; Kerr, M.; Schmidt, J. Overview of PECVD Silicon Nitride for Solar Cells. In Proceedings of the ANZSES 2003—Destination Renewables from Research to Market, Melbourne, Austrilia, 26–29 November 2003.
12. Schmidt, J.; Merkle, A.; Bock, R.; Altermatt, P.P.; Cuevas, A.; Harder, N. P.; Hoex, B.; Van De Sanden, R.; Kessels, E.; Brendel, R. Progress in the surface passivation of silicon solar cells. In Proceedings of the 23rd European Photovoltaic Solar Energy Conference, Valencia, Spain, 1–5 September 2008; pp. 1–5.
13. Dingemans, G.; Kessels, W. Status and prospects of Al2O3-based surface passivation schemes for silicon solar cells. *J. Vac. Sci. Technol. A* **2012**, *30*, 040802. [CrossRef]
14. Hoex, B.; Schmidt, J.; Pohl, P.; Van de Sanden, M.; Kessels, W. Silicon surface passivation by atomic layer deposited Al_2O_3. *J. Appl. Phys.* **2008**, *104*, 044903. [CrossRef]
15. Hoex, B.; Gielis, J.; Van de Sanden, M.; Kessels, W. On the c-Si surface passivation mechanism by the negative-charge-dielectric Al_2O_3. *J. Appl. Phys.* **2008**, *104*, 113703. [CrossRef]

16. Huang, H.; Lv, J.; Bao, Y.; Xuan, R.; Sun, S.; Sneck, S.; Li, S.; Modanese, C.; Savin, H.; Wang, A. 20.8% industrial PERC solar cell: ALD Al_2O_3 rear surface passivation, efficiency loss mechanisms analysis and roadmap to 24%. *Sol. Energy Mater. Sol. Cells* **2017**, *161*, 14–30. [CrossRef]
17. Engelhart, P. Laser processing for high-efficiency silicon solar cells. In Proceedings of the Laser-based Micro- and Nanopackaging and Assembly III, San Jose, CA, USA, 28–29 January 2009; p. 72020S. [CrossRef]
18. Grohe, A.; Knorz, A.; Nekarda, J.; Jäger, U.; Mingirulli, N.; Preu, R. Novel laser technologies for crystalline silicon solar cell production. In Proceedings of the Laser-based Micro- and Nanopackaging and Assembly III, San Jose, CA, USA, 28–29 January 2009; p. 72020P. [CrossRef]
19. Riegel, S.; Mutter, F.; Lauermann, T.; Terheiden, B.; Hahn, G. Review on screen printed metallization on p-type silicon. *Energy Procedia* **2012**, *21*, 14–23. [CrossRef]
20. Canadian Solar Sets a 22.80% Conversion Efficiency World Record for P-Type Large Area Multi-Crystalline Silicon Solar Cell. Available online: http://investors.canadiansolar.com/news-releases?items_per_page=10&page=3 (accessed on 1 June 2020).
21. Kerr, M.J.; Cuevas, A. General parameterization of Auger recombination in crystalline silicon. *J. Appl. Phys.* **2002**, *91*, 2473–2480. [CrossRef]
22. Al-Amin, M.; Grant, N.E.; Pointon, A.I.; Murphy, J.D. Iodine—Ethanol Surface Passivation for Measurement of Millisecond Carrier Lifetimes in Silicon Wafers with Different Crystallographic Orientations. *Phys. Status Solidi A* **2019**, *216*, 1900257. [CrossRef]
23. Aberle, A.G. Surface passivation of crystalline silicon solar cells: A review. *Prog. Photovolt. Res. Appl.* **2000**, *8*, 473–487. [CrossRef]
24. Lee, J.H.; Min, K.H.; Kang, M.G.; Jeong, K.T.; Lee, J.I.; Song, H.-E.; Park, S.; Park, J.-S. Efficiency characteristics of a silicon oxide passivation layer on p-type crystalline silicon solar cell at low illumination. *Curr. Appl. Phys.* **2019**, *19*, 683–689. [CrossRef]

MDPI
St. Alban-Anlage 66
4052 Basel
Switzerland
Tel. +41 61 683 77 34
Fax +41 61 302 89 18
www.mdpi.com

Energies Editorial Office
E-mail: energies@mdpi.com
www.mdpi.com/journal/energies

www.ingramcontent.com/pod-product-compliance
Lightning Source LLC
LaVergne TN
LVHW070611170726
843515LV00004B/49

* 9 7 8 3 0 3 9 4 3 6 2 9 3 *